The Wool Book

THE WOOL BOOK

Margaret Dixon

HAMLYN
London · New York · Sydney · Toronto

ACKNOWLEDGMENTS

Photographs
British Wool Marketing Board, 24, 98–9, 101
Calderdale Museums Services, 55
Craven Herald, Skipton, 12
Colorific!, 33, 56–7
Hebridean Press Service, 48
International Wool Secretariat, 18, 20
Mansell Collection, 14, 21, 23, 29, 41, 46, 50, 58, 76
Michael Plomer, 6, 10–11, 17, 27, 44–5, 68, 86–7, 90–91, 94, 102, 110, 113, 115, 116, 118, 119, 121, 123, 125, 126, 127, 128, 129, 130, 131, 132, 133, 134, 135, 138
Spectrum, 49
University of Reading, Museum of Rural Life, 19
Tom Weir, 52–3
Welsh Folk Museum, 26, 36
WIRA, 16, 96
Lore Youngmark, 95

Line illustrations by the Hayward Art Group

The publishers would like to thank The British Wool Marketing Board, John Lewis Partnership, The Handweavers Studio and Gallery Ltd and Nancy Child for supplying material for photographic purposes; also Mrs Winifred Mason for the use of her house for photography

First published in 1979 by
The Hamlyn Publishing Group Limited
London · New York · Sydney · Toronto
Astronaut House, Feltham, Middlesex, England

Filmset in England by Keyspools Ltd., Golborne, Lancs
in 12 on 14 pt Poliphilus
Printed and bound in Italy

ISBN 600 39426 3

Contents

Overleaf: *A range of attractive and useful garments can be made out of handspun wool.* *See The Patterns on page 111*

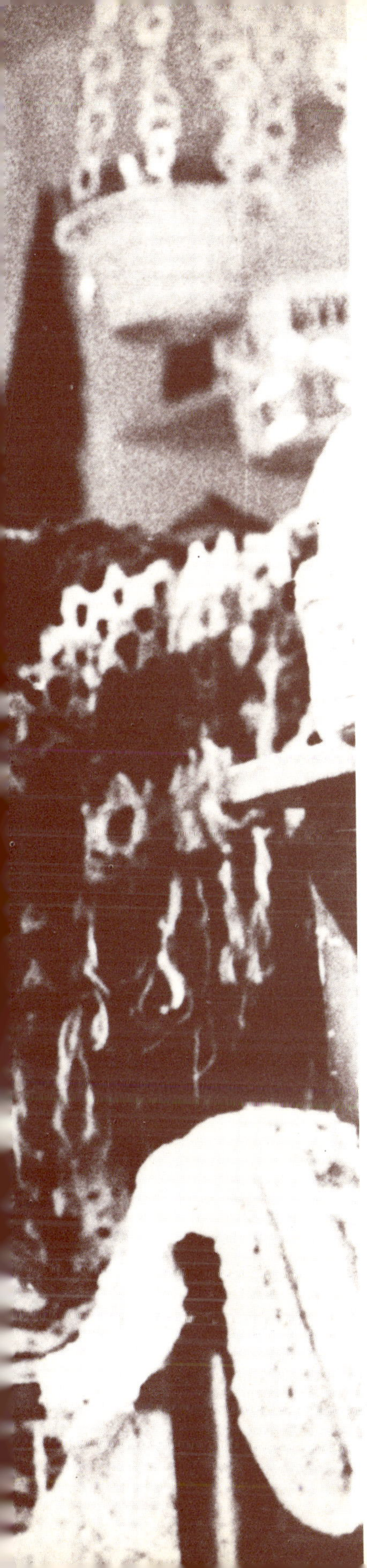

What is Handspinning?

Why Spin?

I suppose that, to some people, suggesting that they should take up handspinning must seem like suggesting that they should turn their backs on two hundred years of history and human progress. It is like implying that the path our civilisation has taken has led us nowhere, necessitating the attempt to go back and start again. Indeed, maybe this does account for the current resurgence of interest in the old crafts, directly reflecting a degree of disillusionment with advanced technology. This could be both because of its impact on the environment and because of the human implications for the people who have to work its machinery. However, just as surely, this is not the most positive reason for pursuing them.

In a world of expanding populations and of escalating expectations in regard to material things, none of these traditional methods could meet our inflated demands for consumer goods nor could they compete commercially with more sophisticated means of manufacturing. So, if we pursue them, we must pursue them either because they offer us a source of enjoyment and interest or, at best, because they offer us the prospect of creating something of outstanding beauty; articles which in their design demonstrate to the full the possibilities of the human imagination, uncompromised by the need to consider profit margins and the practical limitations of machines.

Handspinning does offer these things. First, it may do so by giving us the satisfaction of making something from start to finish, something not implied in the way we organise our factories where the manufacture of articles is divided into its constituent operations, each accomplished by a different individual. Second, it allows us to create something unique in a world where everything tends to be mass-produced and uniform. In addition, since it is a craft with a long and important tradition, it can also assist us to escape for a while from the rigours of modern life by recapturing the romance of a world which, whatever its deficiencies, at least permitted its occupants a greater degree of individuality. Hence, it may awaken in us an interest in history. It is therefore a pastime which at least may be pleasurable and which at best may help us to appreciate what is valuable in our own age and what we have lost in our drive for progress.

The Importance of Spinning

Today, we have many ways of keeping ourselves warm and protecting ourselves against the elements. Not only the means of building substantially draught-free and well-insulated buildings, houses, factories and offices, but also a variety of fuels and of ways of using them, as well as a host of different fibres for the manufacture of fabrics, both natural and synthetic.

Our means of transport, too, we can heat effectively. In a sense then, clothes are probably less important from a functional point of view than they once were, something we require mainly to give supplementary protection and cover up our nakedness: a cosmetic.

Yet clothes were once, besides fire, the only things men had to warm them. The fact that they were of such critical importance is

Eve spinning with a spindle as depicted in the Verislav Bible, 1340

demonstrated by the fact that wool and the other fibres that went to make them commanded such a large share of the market. Thus, Babylon (which, translated, means 'Land of Wool') grew rich by her woollen trade, no doubt in much the same way as middle-eastern states today have grown rich by oil. To the Ancient Greeks too, wool was a byword for wealth; hence, the myth of the Golden Fleece sought by Jason and his Argonauts.

Then, later still, in Medieval England, it was wool again which formed the basis of the wealth both of the country as a whole and of the monasteries, just as it formed the basis of the wealth of Spain during the 16th century. And because it did produce such wealth, states would go to great lengths to protect their woollen industries. England in the 14th and Spain in the 16th centuries passed legislation to prevent the export of rams to which they attributed the superior quality of the wool coming from their livestock.

Then again they would protect it from competition from other fibres. For example, because cotton imported into Europe from India in the 17th century represented such a great threat to wool, statutes made it illegal in England for a woman to wear a cotton garment next to her skin and for a corpse to be buried in anything other than woollen clothing.

Along with wealth, power flowed from wool. In the 14th century, Edward I was able to secure the neutrality of Belgium in his 100 Years War against the French by threatening to cut off supplies of English wool to Flanders which had become over-industrialised and thereby dangerously dependent on England economically. Moreover, in domestic politics too, wool was an instrument of power to kings who could choose to whom they showed their favours by the grant of licences to export wool and raise money to fight wars by imposing taxes on trade in it.

But it was not only in trade and politics that the woollen textile industry is significant. The invention of the spindle for twisting the woollen fibres into yarn is on a level with that of the wheel in technological importance. Indeed it is thought to have preceded the wheel as the first application of the principle of continuous rotary motion.

Moreover, it has had much more of an impact on our lives today than many of us realise because it was the woollen industry in England that first saw the introduction of the Factory System in the 17th century, which laid the foundations for the way in which men would organise their fellows to work both during and after the Industrial Revolution.

What is Wool?

At its most basic, wool is a fibre composed of a protein called keratin. This, in itself an exceedingly complex substance, is also arranged in a remarkably complicated way to form a single woollen fibre. Even then, this arrangement varies from one breed of sheep to another, accounting in part for the fact that the characteristics of fleeces vary from breed to breed, some being long in the length of the fibres (sometimes as much as 40 cm (16 in) in the Lincoln Longwool) some being short (sometimes as short as 6 cm ($2\frac{1}{2}$ in) in the Merino).

Again, some wools are more lustrous than others and this is due to differences in the structure of their fibres. Under a microscope, the woollen fibre most resembles a tall stack of plant pots, known as epithelial scales. In some wools, these overlap each other more than in others. Thus, a wool in which they overlap each other by no more than a third of their full depth is less scaly, appears more lustrous and is therefore described in the woollen trade as a lustre wool. Such a wool is that of the Lincoln. That of the Merino, on the other hand, whose fibres have scales which overlap each other by two-thirds of their full depth is much more scaly and almost dull in appearance by comparison.

Again, some wools are coarser than others. The Merino, which gives the finest wool, has fibres measuring as little as 23 microns whereas the coarsest wools measure as much as 150 microns in diameter. Yet in a single fleece, not all its fibres are of equal coarseness.

To begin with, in all sheep, their woollen fibres grow from two different kinds of follicles, called primary and secondary. Those which grow from the secondary follicles are much finer than those which grow from the primary. The proportion of primary to secondary follicles in a fleece varies from one sheep to another. Thus in some fleeces, the proportion of secondary to primary follicles is of the order of three to one. This makes the whole fleece generally coarse in character. In others, however, such as the Merino, the ratio is in the region of twenty secondary to one primary, accounting for its extremely fine overall appearance.

The fineness of a sheep's wool may vary from one part of its body to another. That which comes from its belly is finer than that which comes from its hindquarters or back. The wool from the neck, too, may be extremely fine. In fact, some of the most remarkable examples of handspinning are Shetland shawls spun so fine that they can be drawn through a wedding ring. These are made from wool plucked (or 'rooed') from the neck of the Shetland.

Wool fibres magnified 400 times show clearly the arrangement of epithelial scales

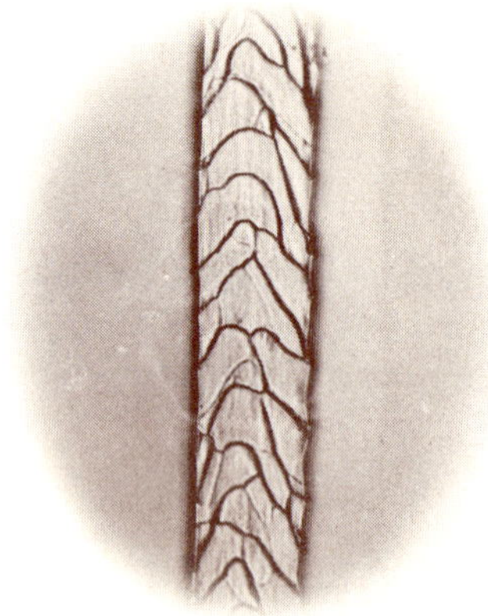

Fine wool from a Shetland

Coarser wool from a Southdown

Opposite: A Shetland shawl spun so fine it will pass through a wedding ring

As to the other characteristics of wool, one of these is called crimp. This refers to the waviness of the individual fibres and is sometimes measured in terms of crimps per inch. Some wools are very crimpy, others not so. This is a useful attribute in spinning because the corrugations of the different fibres, as they are drawn out, twist around each other, making the resulting yarn stronger. Indeed, if one should ever try to spin a fibre without any crimp, it may be advisable to mix this with a quantity of crimpy woollen fibres.

Another feature is the grease to be found in wool. This is a natural secretion of the sebaceous glands situated in the sheep's skin close to its wool follicles. As long as it has not been washed out by rain or dried out by sun and as long as the fleece is not too dirty so that it has to be cleansed of both natural grease and dirt before spinning, this grease too is useful to the handspinner for it helps the fibres being spun to cling together.

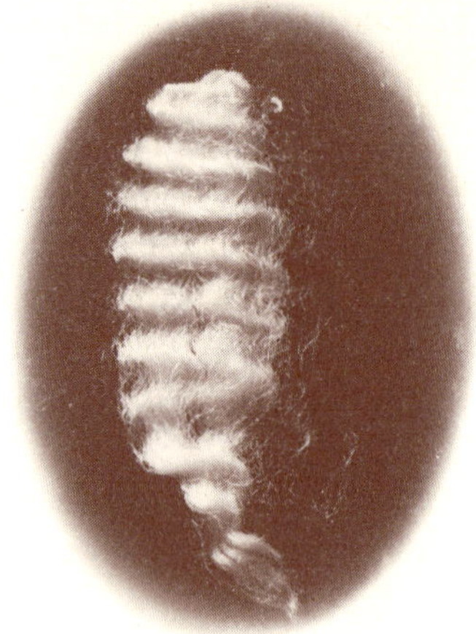

Crimpy fibres are easier to spin

The Origins of Wool

Primitive sheep such as the Moufflon, the Argali and the Urial, had both a coarse outercoat, more like hair than wool, and a soft undercoat, as fine as that produced by any Merino. Their fleeces were in fact not dissimilar in some respects to those still found on some hill breeds of sheep such as the Scottish Blackface and the Soay which is still found on the Hebridean island of St Kilda.

As to colour too, individually the sheep might offer a range of different browns and greys as well as black and white, something which is now only offered by a few breeds, notably the Jacob or piebald sheep, the Shetland and the Orkney. However, despite the farmer's best efforts, even white breeds of sheep occasionally throw up black or brown individuals and some people, particularly some handspinners, have seized on these to breed from in the hope of producing a consistent line of black and brown animals.

Yet clearly this is not the general picture, particularly as it applies to commercial sheep breeding. From earliest times, it has been the sheep farmer's prime aim to get both white wool (because it dyes easily and offers the widest range of possibilities to the textile designer) and whole flocks of sheep whose wool is of a consistent quality. Consequently, most wool that is offered today is white and it is possible to say, within certain narrow limits, that a particular breed of sheep will produce a particular type of wool, a luxury that was by no means available to the ancients.

Sheep from the Shetland and Orkney Islands

Top: A ram from the Isle of Enhallow

Right: A ewe from the Isle of Ronsay

Left: The lamb is a cross with Cheviot

Yet even this does not explain why certain breeds of sheep are found in particular parts of the world. This can only be understood by reference to history and environment.

As to the latter, obviously not all climates and terrains are suitable for the rearing of all kinds of sheep. On high, exposed land in cold climates, a tough, resilient breed such as the Rough Fell or Scottish Blackface proves most profitable. On low-lying, marshy land such as that in much of New Zealand, a breed such as the Romney Marsh which was bred especially to thrive in similar conditions in South-East England, is the breed which finds the most favour.

History offers even more by way of explanation. For as adventurous men roamed the world in search of new lands to colonise, they took their animals with them. This is the reason why many of the breeds of sheep found in, say, New Zealand, Australia and North America are the same as those found in Great Britain.

Yet, since the 16th century, there has been another major influence at work in determining the dispersal of sheep about the world. This is the extraordinary wool-bearing quality of the Merino. Originally reared in Spain, the product of the Laodicean and Tarentine sheep brought there by the Romans, its wool was always so desirable that, wherever it would thrive, it was taken up and quickly took precedence over other breeds. It has maintained this position at least in those countries where the primary objective of sheep-farming is still to get wool rather than mutton. This is the case in Australia for example. It was also the case in England at least until the 18th century.

It occurred in New Zealand too, until the introduction of refrigeration in 1882 brought a change of commercial emphasis. Since then, the Merino has fulfilled another role in such parts of the world. In New Zealand, it has been crossed with other breeds such as the Leicester and the Lincoln to produce the Corriedale and, elsewhere, with still more breeds to add to whatever desirable characteristics these were thought to possess, its ability to produce good wool.

This is a continuing process and, no doubt, in years to come, still more new breeds will be created, perhaps with the help of the Merino, perhaps with the help of some of its younger cousins; each suited to the markets of the world and their changing circumstances.

Merino rams

The Principle of Spinning

Spinning in ancient Egypt as shown on a fresco at Beni Hassan

Spinning itself, whether done with wool or with any other fibre (flax, silk or any synthetic fibre for example) consists of two actions. Firstly drawing out the individual fibres from a mass and secondly twisting them together to form a yarn. The thickness of the resulting yarn obviously depends on how many of the fibres are drawn out at once and twisted.

Of all the fibres, wool is probably the easiest to spin because it already has all the qualities necessary to make them cling together; the grease, crimp and that scaliness which causes them to mesh into each other – unlike, say, flax which the spinner must wet with her fingers as she is spinning it in order to mould it into its twisted shape. But because different wools possess these attributes in varying degrees, some wools are easier to spin than others. Besides, the fibres of some wools are longer than those of others. Therefore, for the purposes of spinning by hand, a wool whose fibres have a staple (a length) of between 10 cm (4 in) and 15 cm (6 in) is easier to spin than one with a staple of, say, only 5 cm (2 in).

As to the discovery of this principle of spinning, however, it is perhaps not too difficult to imagine how man found out how to make yarn. What may be more difficult, though, is to picture what he thought he would do with the finished yarn. Yet maybe he had already discovered weaving when building primitive dwellings with reeds and rushes.

But since yarn can be made without instruments of any kind (except the thumb and forefinger) simply by drawing out the fibres by pulling them apart with both hands and twisting them; no doubt, like so many other human discoveries, that of how to make yarn was an accident. Originally this would be a slow process for the first difficulty the spinner would encounter is the tendency of the fibres to untwist themselves. Therefore, the next stage in the development of spinning technology was probably the use of a stick on which to wind the spun yarn.

Such a stick formed the basis of the simple spindle. It was already well established in Neolithic times and still the only instrument in use for spinning until, at the earliest, the 12th century AD. For in principle, the spindle is no more than a stick with a weight attached at some point to give it momentum when spun either suspended from the end of a thread or with one end resting on the ground like a spinning top. But of course none of this would ever have happened if man had not first had wool and a means of separating it from the

Possible Development of Sheep Breeds and their Fleece Types

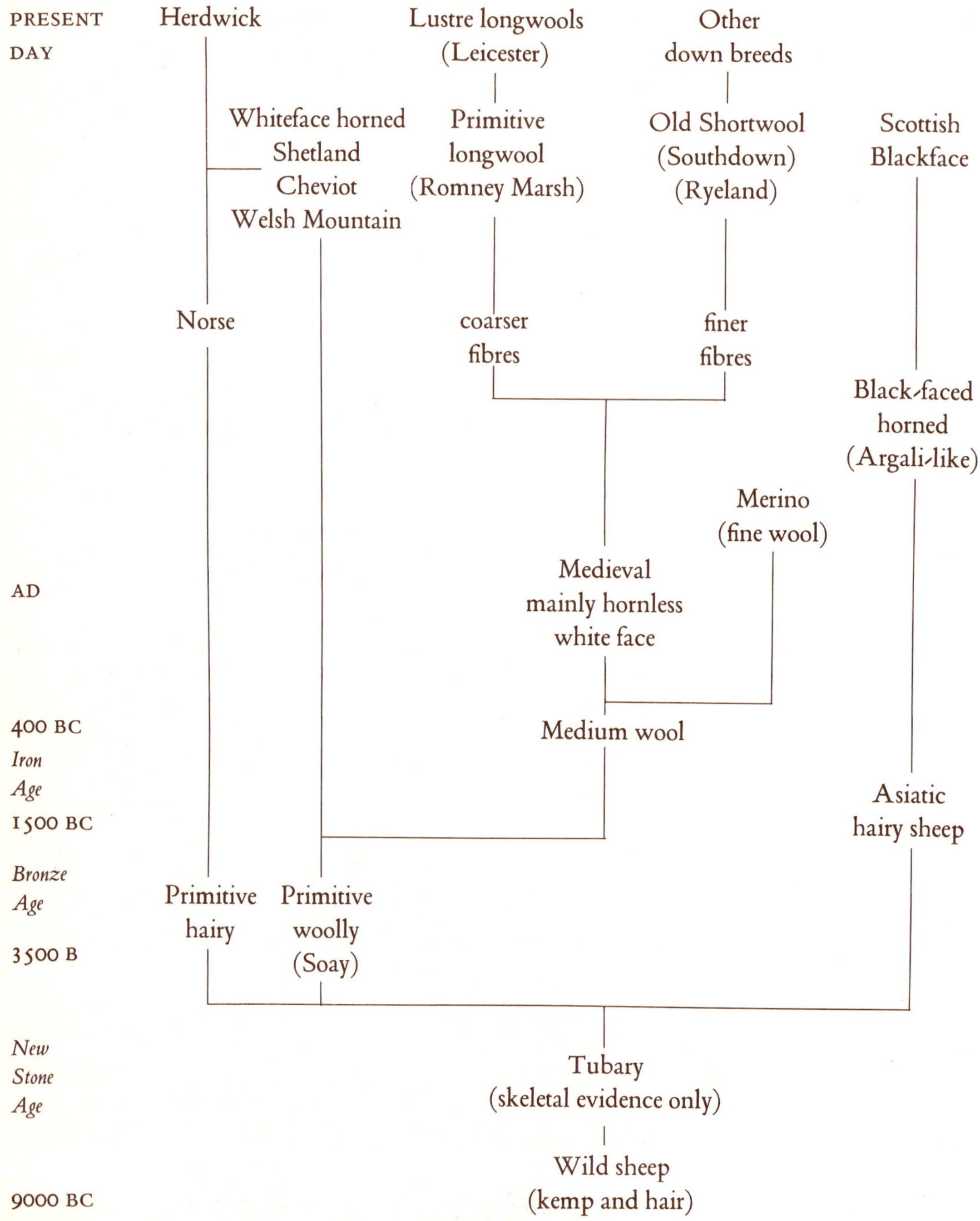

sheep. This was no problem as long as moulting was natural as it was in primitive sheep. All he therefore had to do was collect it where it fell.

Yet as men domesticated the animal and interfered with its breeding, this characteristic for the most part disappeared. They had then to invent some instrument to remove the wool. This resulted in the shears which are still used, having been employed consistently for this purpose and virtually unchanged since the days of the Romans. Their general use was only brought to an end by the introduction of electric shears.

The evidence of the way in which spinning wool developed and when it began, comes from the investigation of archeological sites. Many spindle whorls (the weights, usually in the shape of a disc, which were attached to the spindle) have been found, many of them decorated and carved in attractive designs. Apart from these, samples of woollen cloth have been found which not only give us an insight into the methods employed by stone age men in weaving, but also enable us to deduce what their primitive sheep must have been like by analysis of the fibres.

Spinning with a spindle using a distaff to hold the fibres

First Steps in Spinning

The Natural State of Fleeces

The first obstacle to the spinning of wool may be the condition of the fleece whose fibres are to be spun. Ideally, such a fleece is relatively clean, moderately greasy and not tangled. If it is in this state it is possible, at least for the practised spinner, to process it into yarn without further ado. However, since this is by no means always the case, some preliminary treatment of the fleece may be necessary. In the first place, if it is tangled, its fibres will need some separation.

In early times, the spinner, before attempting to spin the wool, would draw it out from the fleece and roll it against the thigh into a thick band of wool called a roving. A Greek vase of about 560 BC shows such a spinner engaged in doing this with his thigh protected by an unglazed tile, shaped to fit around his thigh and knee. This was called an epinetron. The roving would then be wound around the top of a length of wood which was held either in the hand or in a belt. Both the Greeks and the Romans used such a means of storing the supply of fibres to be spun; a primitive form of the distaff which is attached to many spinning wheels.

But even so, such a crude form of preparing the fibres is not always adequate for the eventual purpose for which they are intended. Plainly too, the spinner still has to contend with a degree of tangling and fibres which will not draw apart easily. For this reason, a means of achieving a higher level of organisation of the fibres was invented.

The first of these was the Roman carduus which was a cross of wood on which were mounted teasels. By drawing these across the wool, it is thought that the spinner did much to disentangle the fibres prior to spinning though there is some controversy about this. There are those who argue that the carduus was an instrument used only for raising the nap on woven cloth. However, if it was used for disentangling wool, no doubt it made the spinner's eventual task much easier. Still another means of disentangling the fibres, however, was by beating them with a gut-stringed bow, thereby vibrating them. Yet though this continued alongside the development of such inventions as the carduus, neither could compete with the much more efficient carder.

This latter instrument was and is a much more effective tool. Carders are flat or curved rectangles of wood, one surface of which is covered with leather into which are set rows and rows of short, fine, bent wire teeth, called scratchcard; they are also equipped with handles. Yet even these only suffice to untangle the wool and attain an even dispersal of the fibres in the mass of the wool they are set to deal with (called a rolag). They still leave short fibres in the rolag and these may lie across the general direction of the longer ones, causing the resulting yarn to be fluffy. This is adequate for the formation of yarn which is to be made into some kinds of cloth and for knitting and crochet yarns but not for that of yarn to be made into worsted. For here is a distinction of great importance in the woollen textile industry – that between woollen and worsted.

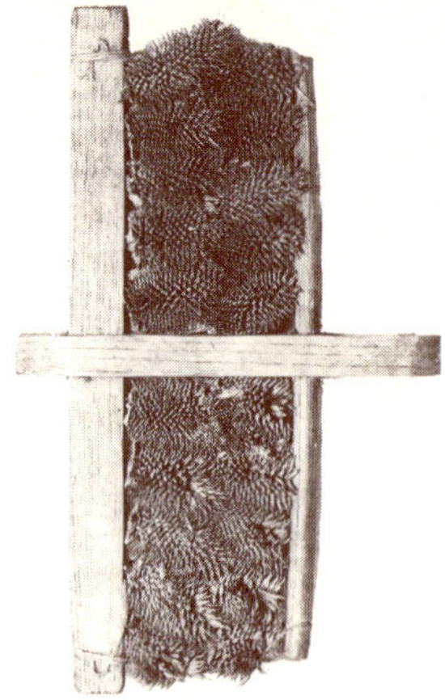

Teasel carder

Visually, it is easy enough to distinguish between the two as woollen cloth is by no means as smooth as worsted. However, the real distinction lies in the way the fibres are prepared for each prior to spinning. For woollen yarn, the fibres are only carded. For worsted they are combed with combs consisting of long, pointed prongs, set into wood in several rows. These succeed not only in disentangling the fibres and in arranging those which remain into a parallel formation, but also in removing all the short fibres which might eventually lead to the production of a rougher length of cloth. These are called noils. Combing was always heavy and skilled work, a craft on its own. Before the introduction of mechanical combs it was the work of men who travelled about the country offering this service.

As to the cleanliness of the wool to be spun, in some respects this may present few problems. Pieces of dung stuck to the hind-quarters of the fleece may simply be cut off. Burrs and foliage too, may be removed by hand, whereas smaller objects such as grass-seed can be removed by carding. On the other hand, an excessively dirty fleece, one from a sheep kept on peaty soil for example, may need washing before spinning. Naturally this removes the grease from the wool too and therefore, a substitute in the form of, say, olive oil, has to be introduced to make the wool once more spinnable.

Preparing the Wool to Spin

It is possible for the handspinner to acquire already prepared wool for handspinning. Such quantities of wool are called 'tops' and are fibres which have been prepared by machine for spinning on commercial spinning machines. However, from the point of view of handspinning, there seems to be little purpose in taking this short cut to the acquisition of a supply of suitable fibres since at least one of the appealing characteristics of handspun wool is its unevenness, an unevenness which is often deliberately contrived to give it a particular texture. Handspinning with tops, on the other hand, reduces the degree of this unevenness and consequently the degree of individuality of the finished garment.

So, if you accept this argument and start your handspinning from the raw wool, the opposite extreme to using machine-carded wool is to prepare it for spinning by merely teasing. This is simply separating

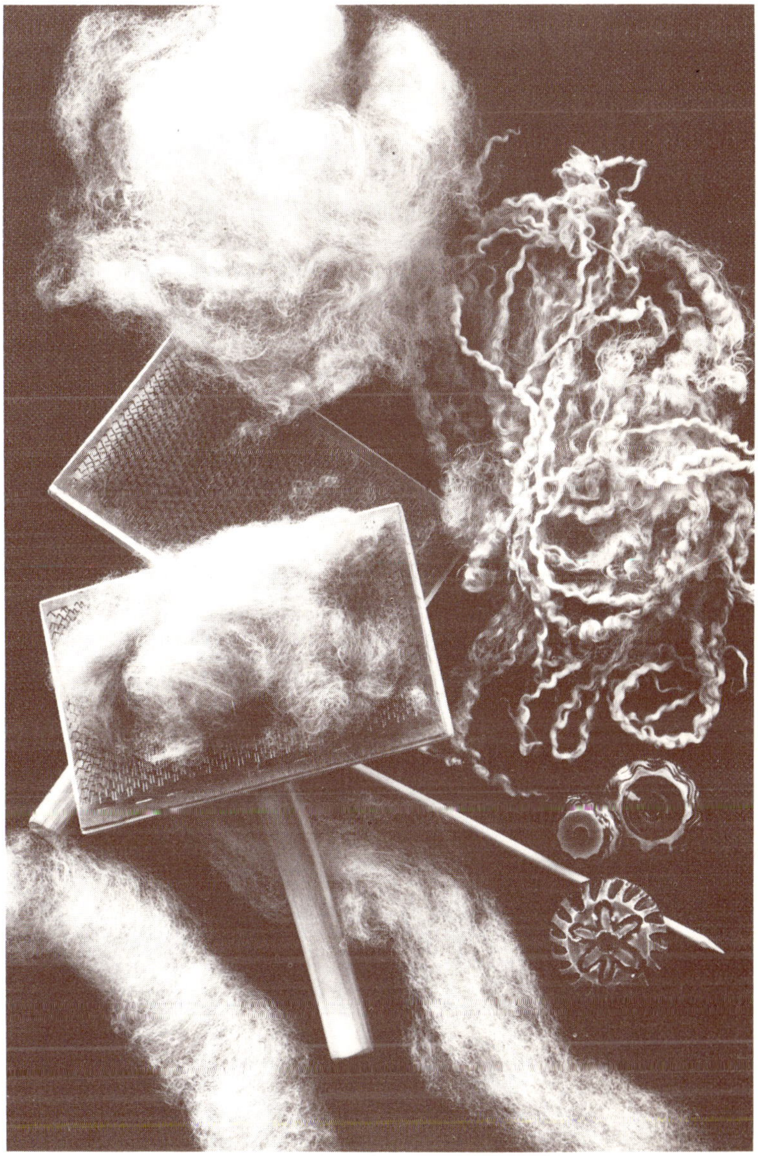

Below the carders lay two rolags ready for spinning. The spindle whorls are modern copies of those excavated from Neolithic sites

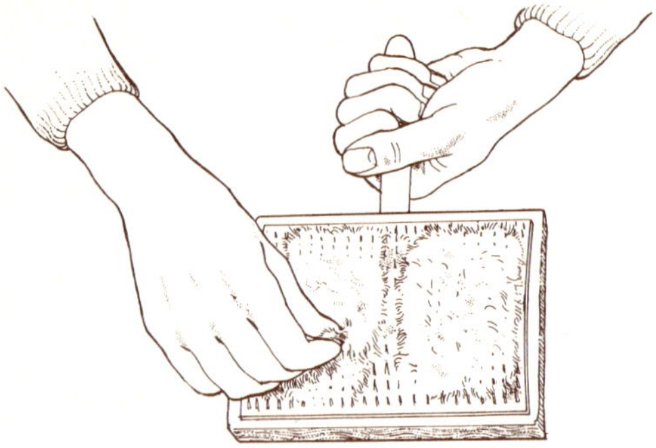

1. Lay the left hand carder on your lap and place some wool across the teeth

2. Draw the right hand carder firmly across the left hand carder. Repeat this several times until the fibres lie parallel and are evenly distributed between right and left hand carder

3. Transfer the fibres remaining on the left hand carder on to the right hand carder, by pushing the teeth of the left hand carder across those of the other

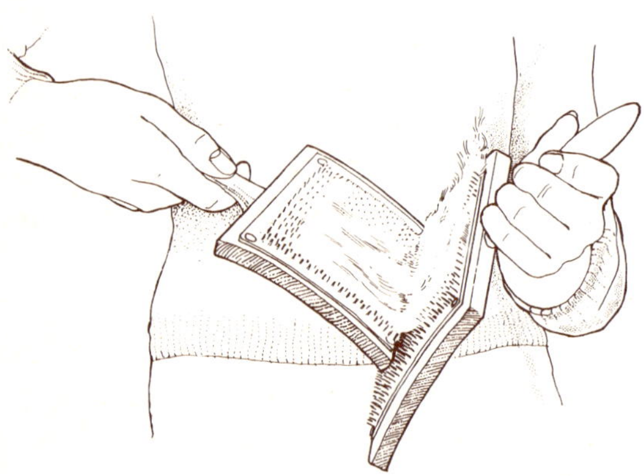

4. Transfer the sheet of wool back to the left hand carder by repeating and reversing this process

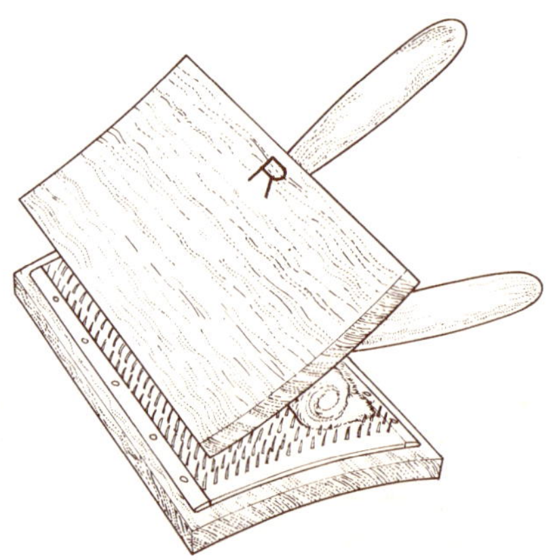

5. Stroke the sheet of wool on the left hand carder with the teeth of the right hand carder to form a roll

the fibres by gently pulling them apart with your fingers, removing dirt and vegetable matter as you go. At one stage further on, however, the process of carding produces an even more orderly arrangement of fibres.

Carding

Ideally you should mark your carders for your left and right hands and use them accordingly. Lay the left hand carder flat on your lap with its teeth facing upwards and its handle away from you. Then gripping the handle firmly in your left hand to keep it steady, distribute a small amount of wool in strands over the whole of the toothed surface of this carder. Do this by stroking the fibres on to the teeth across its entire width. Then, taking the right hand carder, use it to comb the fibres in the left hand carder until they lie parallel. In doing this, you will evenly distribute the wool between both carders.

When all the fibres are straight and parallel and all of the noils and impurities have been removed, make the wool into a rolag as follows. First transfer the fibres in the left hand carder to those of the right hand one by pushing the teeth of the left hand carder against those of the other. At this stage, unlike before, the handles of both carders are facing in the same direction. Then transfer the fibres on the teeth of the right hand carder back on to those of the other by repeating and reversing this process. All the fibres now lie on the left hand carder. Gently remove them either by hand or preferably by stroking them with the teeth of the right hand carder in such a way that they form themselves into a roll.

Combing

The wool comber

As already mentioned, the main difference between carded and combed fibres is that carded fibres are essentially an orderly tangle which includes at least a proportion of short fibres. Combed fibres, on the other hand, are consistently long and more or less completely disentangled fibres lying parallel. This accounts for the superior quality of worsted.

However, there is another difference between woollen and worsted. This lies in the way the fibres are drawn from the fibre supply as they are spun. Woollen yarn is drawn from the end of the rolag in which the fibres are all lying at right angles to the length of the rolag and hence to the direction from which they are being drawn. Worsted yarn, however, is drawn from a roving (equivalent to the rolag) whose fibres are parallel to its length and hence are drawn off in the same direction as they lie in the roving.

As to how combing can be done, usually, in the past, it was accomplished with two long-tined combs. One of these was securely fixed to a post or something similar. A wad of fibres would then be pulled through the tines over and over again, leaving all the noils behind on the comb.

This could be achieved with only one comb but again it was usual to use two, in much the same way as carders, drawing the one across the other. Even so, one was still fixed, the other free. The free one was generally kept warm on a stove when not in use because its warmth when drawn through the wool, loosened the natural grease in the wool and made the fibres slide more easily past each other.

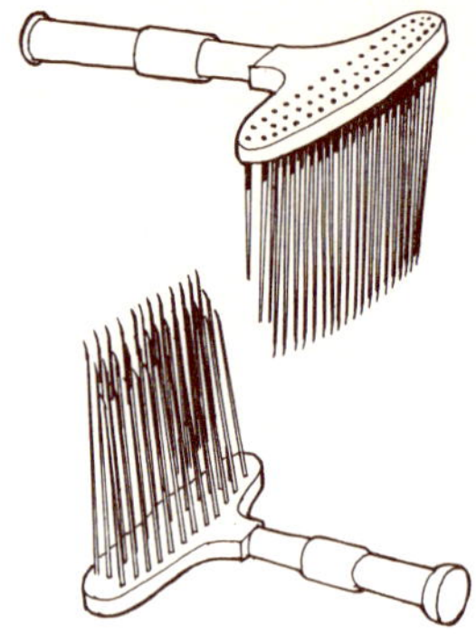

A pair of combs

Washing

If a fleece does need washing before spinning, either because it is very dirty or because it contains too much grease (lanolin) you can wash it. But unless there are other reasons for wanting to remove both dirt and grease, then attempt to wash it in such a way that you remove the dirt without removing the grease as the latter is a valuable aid to spinning. Fuller details of how to wash a fleece either for dyeing before spinning or just to render it clean enough to spin, are given in Chapter 5.

However, because of the bulkiness of a fleece, it is advisable before you do anything else to sort it into different grades of wool first and wash each batch separately. Later, when they are dry, if you want to obtain an even quality of wool throughout the fleece, you can do this by carding together batches of coarser wool with batches of finer wool. It is always easier to sort a fleece into different grades of wool while it is 'in the grease'.

Sorting a fleece

When a fleece is clipped from a sheep, it is done in such a way that the shearer ends up with a fleece which is in one piece and which, when laid out, resembles the shape of the animal. It is then the common practice to roll this up, starting at the tail or britch end and tying it round with the long narrow band of fleece which comes from its neck. Thus, when you are presented with a fleece, it will usually be in this rolled up form. To sort it, you should first untie the neckband from around the bulk of the fleece and then unroll this to its full length.

When sorting, first apply your attention to the poorer qualities of wool. These will usually be from around the legs, britch, belly area

Opposite: The fleece ready for sorting

The Fleece

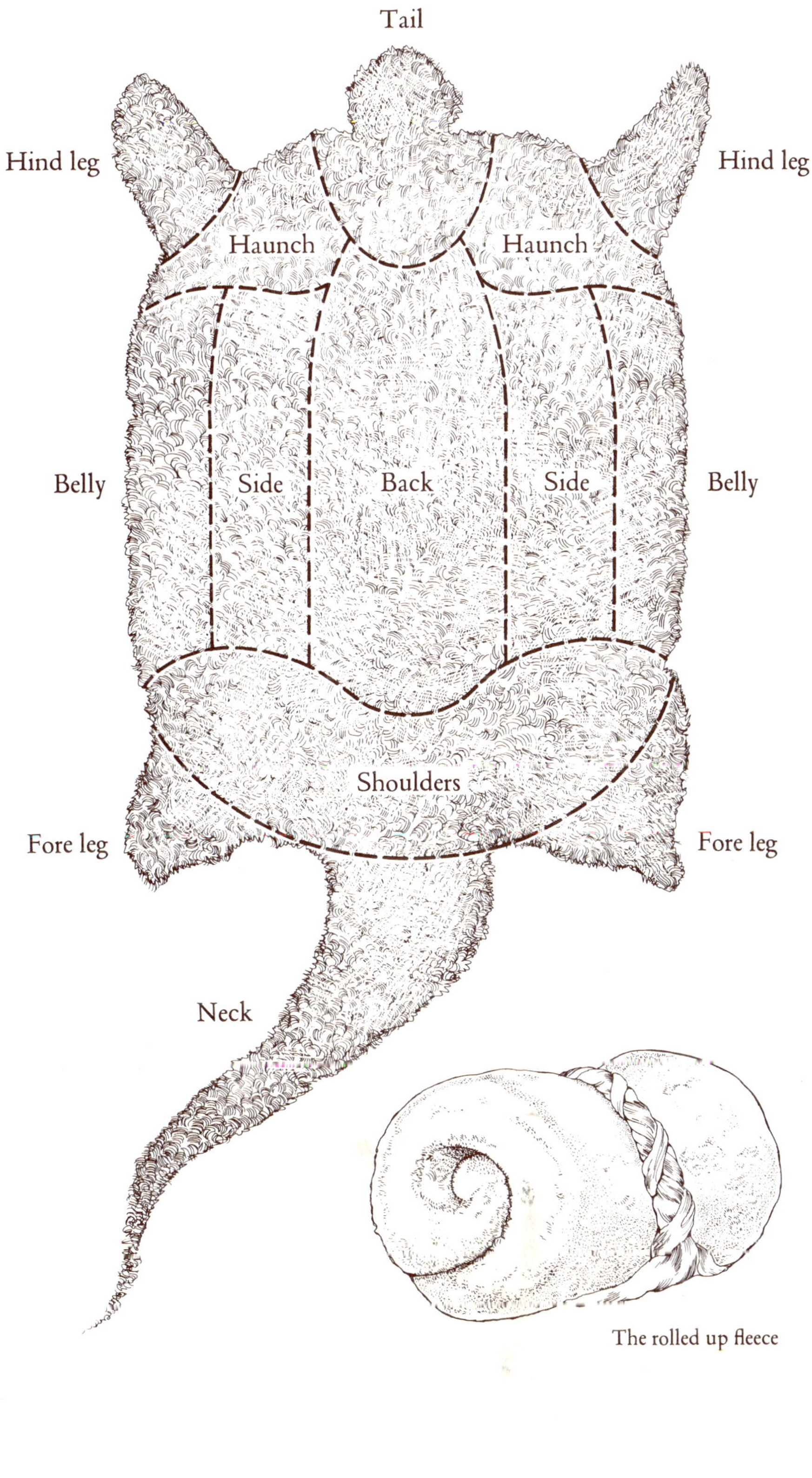

The rolled up fleece

and head and may be unsuitable for spinning, in which case remove them and discard them altogether. Then turn to the rest of the fleece.

Of this, the best wool will probably come from the shoulders and sides though the latter may not be quite as good. The neck too, may provide soft wool as may the belly, though that from the latter often proves too dirty to use. Wool from the back of the animal is likely to be dry, with a tendency to be brittle because of its exposure to the weather. Finally, the wool from the haunch or britch part of the animal, as well as that from the upper part of the legs, is usually of a coarser quality than the rest. When you have decided which parts of the fleece are which, pull it apart, separating them from each other, and then deal with each section accordingly.

How to Spin with a Spindle

There are several different kinds of spindle although they all work on the same principle. Some of them are still in general use in underdeveloped countries such as Guatemala, Peru and the Sudan and are also employed by the Tuareg of the Sahara and the North American Indians. However, the one in everyday use in Europe before the invention of the spinning wheel, consisted of a disc, either of wood or clay, with a hole in the middle, through which passed a length of dowel. Such spindles were suspended from a length of yarn and spun in this position, the fibre supply being held in one hand or on a distaff, and the fibres to be spun being drawn out from the fibre supply with the other.

Another type of spindle, though, was generally larger and employed like a spinning top, with its lower end resting on the ground. In both cases, however, when the yarn is spun to a manageable length, it is wound on to a dowel and the process of drafting out the fibres and twisting them begins again.

As already stated, the invention of the spindle was, in technological terms, an extremely successful one, being employed from Neolithic times even to the present day in some societies. Indeed, even when the first crude versions of spinning wheel were introduced into Europe in the 12th century, there was resistance to their use by guilds of spinners and weavers so that the statutes of Abbeville and Spires, for example, of 1280 and 1288 expressly forbade the use of the spinning wheel for the manufacture of warp threads because it was thought that the spindle could manufacture threads which were much stronger.

Spinning on a supported spindle

Spin the spindle like a top with one hand while drawing out the fibres with the other. This can still be seen going on today in parts of South America

Spinning on the drop spindle

A spindle is easy enough to make. Indeed you can even make one with a length of dowel or a pencil weighted with an apple or a potato. However, a much more effective one would be manufactured out of an 18 cm (7 in) length of dowel and a disc of plywood with a hole in the middle. Better still, one of the same general shape, turned on a lathe, would achieve a spindle of superior balance and one less likely to oscillate, an undesirable characteristic.

In making it, however, there are three main points to consider – the length of the fibres to be spun; the eventual thickness of the yarn and the weight of the spindle. For if it is heavy, if the fibres to be spun are short and if you are planning to spin a fine yarn, you will find that the weight of the spindle causes the yarn continually to break as you spin. Plainly then, for spinning fine yarn, you need a light spindle with a light weight or disc attached to it, and for a thick yarn, you can afford to use one which is heavier.

Spinning on the Drop Spindle

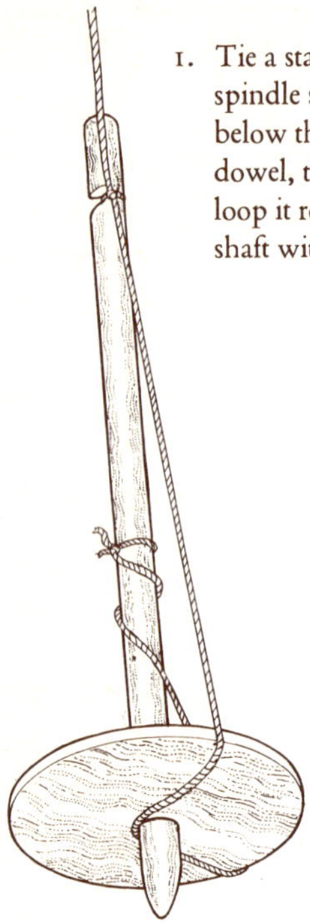

1. Tie a starter yarn to the spindle shaft; take it down below the disc and round the dowel, then bring it up and loop it round the top of the shaft with a half hitch

2. Draw out some fibres from the rolag using thumb and forefinger of each hand. When commencing to spin these should overlap the starter yarn by about 20 cm (8 in)

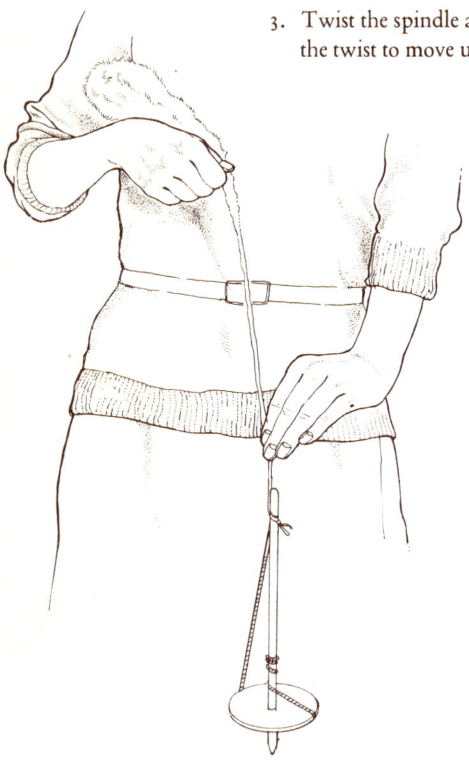

3. Twist the spindle and allow the twist to move up the fibres

4. Allow the spindle to drop towards the floor while you draw more fibres and control the twist with your fingers. When the spindle reaches the floor detach the yarn from the top and bottom of the spindle shaft, wind the newly spun wool around the shaft above the whorl to form a cone. Then leaving a sufficient length of new yarn to tie round the bottom and top of the spindle shaft as before. Repeat the process

As to the method, you can start spinning with a length of yarn already spun. Tie it to the dowel just above the disc, run it over the edge of the disc and round the dowel below the disc and then back over the edge of the disc to the top of the spindle where you should loop it round the top of the spindle with a half-hitch. Then, suspending the spindle from the yarn remaining, draw out a length of fibres from the end of your rolag, about 30 cm (12 in) in length, overlapping these with the previously spun yarn. Then rotate the spindle.

As the spindle spins it imparts twist to the unspun fibres. It is necessary to control this twist as it runs up these fibres by gripping them with the thumb and forefinger of the right hand as you draw out more fibres from the rolag. Otherwise the twist will run up the unspun fibres as far as the rolag and prevent you from drawing out more fibres.

This area between the thumb and forefinger of the right hand and the end of the rolag or fibre supply is triangular in shape and is called the drafting area. Rotate the spindle in a clockwise direction. This imparts what is known as a Z twist to the resulting yarn as opposed to an S twist which, though sometimes employed in the manufacture of single-ply yarns, is only employed for special purposes.

Spinning on a supported spindle

This differs from spinning with the suspended spindle as this spindle is in contact with the hand all the time it is rotating. The spinner, in a sitting position, either rolls the long length of dowel against the thigh from hip to knee to impart a clockwise twist or, holding the top of the spindle between thumb and forefinger, turns it like a spinning top. Generally, this is a much less efficient method of spinning than with the suspended spindle but at least the spinner can remain seated while working.

However, in either case, two points are worth noting: try to keep the spindle rotating and your drafting of the fibres at a constant pace. Spinning in this way should eventually become a regular, easy motion and indeed, in the past, was practised in conjunction with other activities. Thus a woman would spin on her way to the well to fetch water or whilst sitting on the back of a camel.

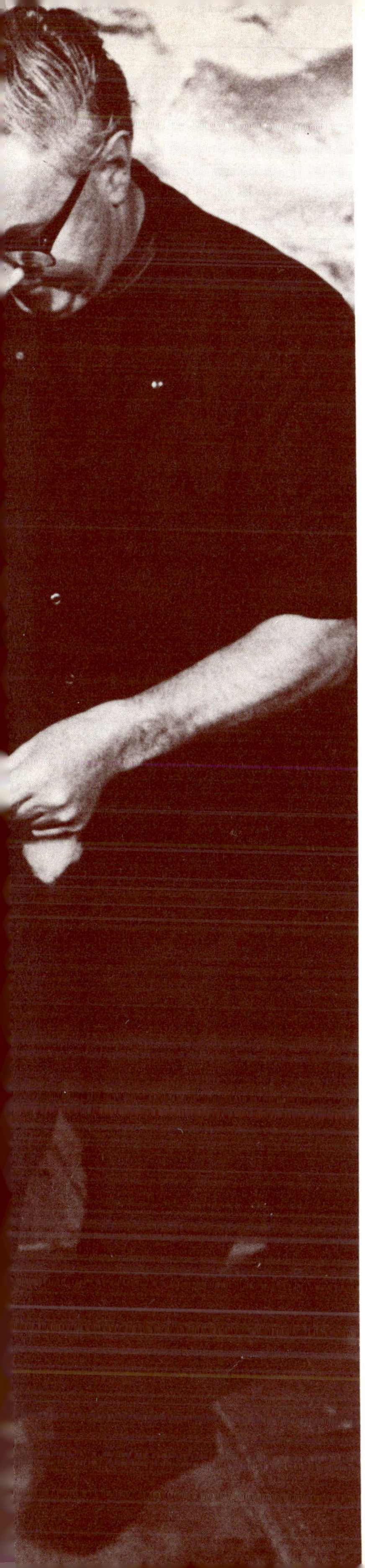

Spinning on the Great Wheel

The Origins of the Great Wheel

The great wheel, sometimes called the walking wheel, the muckle wheel or the spindle wheel, had its origins in China. There, however, in the form that it existed, it was not used for spinning but for unwinding silk from cocoons. Yet from China the principle that underlay this silk-unwinding device travelled to India where, called a charka, it was adapted for use by spinners as the equivalent of the European spinning wheel. This instrument then found its way into Europe and eventually to Britain, it is thought, by way of Holland.

In principle, all that had happened was that the simple spindle had been attached to a post in such a way that it could still rotate but in a horizontal position. The whorl now became a pulley around which a driving band ran to turn the spindle shaft. This driving band itself derived its power from a wheel much larger than the pulley/whorl (something of the order of 20 times larger, in fact) on the spindle shaft. This larger wheel was turned by hand and connected to the pulley/whorl by the driving band. By means of such gearing then, the spinner could now rotate the spindle shaft at a much greater speed than ever before and, because of the weight of the larger wheel, giving it momentum, keep the spindle shaft turning for much longer.

As to the date at which such an invention appeared, it has been suggested that, in India at least, it first saw the light of day in about 500 BC. However, an illustration in Joseph Needham's book, *Science and Civilisation in China*, is thought to be more or less contemporary with the appearance of the spindle wheel in Europe in the 13th century. However, it did appear in different guises in different parts of the world. Though it was usually mounted on a plank called the bed or saddle, in the Faeroe Islands for example, the device was sometimes found attached to the fabric of a building.

Moreover, though the driving wheel itself was sometimes constructed in a manner similar to the way in which a wheelwright would manufacture a wheel for a wagon, with a number of spokes fixed at one end into a hub and attached at the other to a broad wooden rim, it was not uncommon for the pattern of such a wheel to be much cruder. For example, though the spokes were still set into a hub, the rim would be fabricated by bending thin struts of wood to the required curvature, constructing two thin hoops in this way and joining them together at a distance of about 8 cm (3 in), then giving them the substance to carry the driving band and hold the spokes by binding the space between them in a herring-bone pattern with strips of leather. These are known as hoop-rimmed spinning wheels.

However, at its most crude, such a wheel may consist of nothing more than two six-pointed star-shapes made out of three lengths of wood, joined at their centres, revolving about these centres but with the points of one star set so that they divide equally the distances between the points of the other. These distances were then kept constant by joining the points of both stars with cord which, because the two stars were set at a distance apart, zig-zagged between the points around the full circumference. It was on these cords that the driving band rested and moved as the stars were turned with a handle.

Again, the method of fixing the driving wheel also varied. Thus, it was sometimes attached at its hub to two posts, sometimes to one and sometimes, as in the Faeroe Islands, to the interior wall of a dwelling.

As to the spindle shaft and its pulley/whorl, however, this was invariably attached to a single upright post, set into a plank of wood (the saddle) in such a way that the spindle shaft was fixed at right angles to the driving wheel. Many methods of attaching the spindle shaft to such a post have been employed but perhaps the simplest were with leather thongs or with plaited straw. However, though it might be thought that leather would be the most favoured, this was not so, as it required regular treatment in order to remain relatively frictionless whereas straw has a naturally slippery surface. Such means of attaching the spindle could also regulate the tension of the driving band. As long as this remained taut, it would pull the spindle shaft free of the post to which it was attached so that the only friction it would encounter was from the thongs of straw or leather.

The size of the loops created by these thongs could be adjusted and thereby either pull the spindle shaft and pulley whorl further back towards the post which held them or else release these in the direction of the driving wheel. Even so, other methods of adjusting the tension of the driving band were employed such as wedges in the mortice, which held the post to which the spindle was attached, the shape and size of which could affect the angle at which this post leaned backwards.

As to the size of the driving wheel, this too varied somewhat. Perhaps the American walking wheel had the one of the largest diameter, being in excess of 1·25 m (48 in), with an overall height of

The Great Wheel

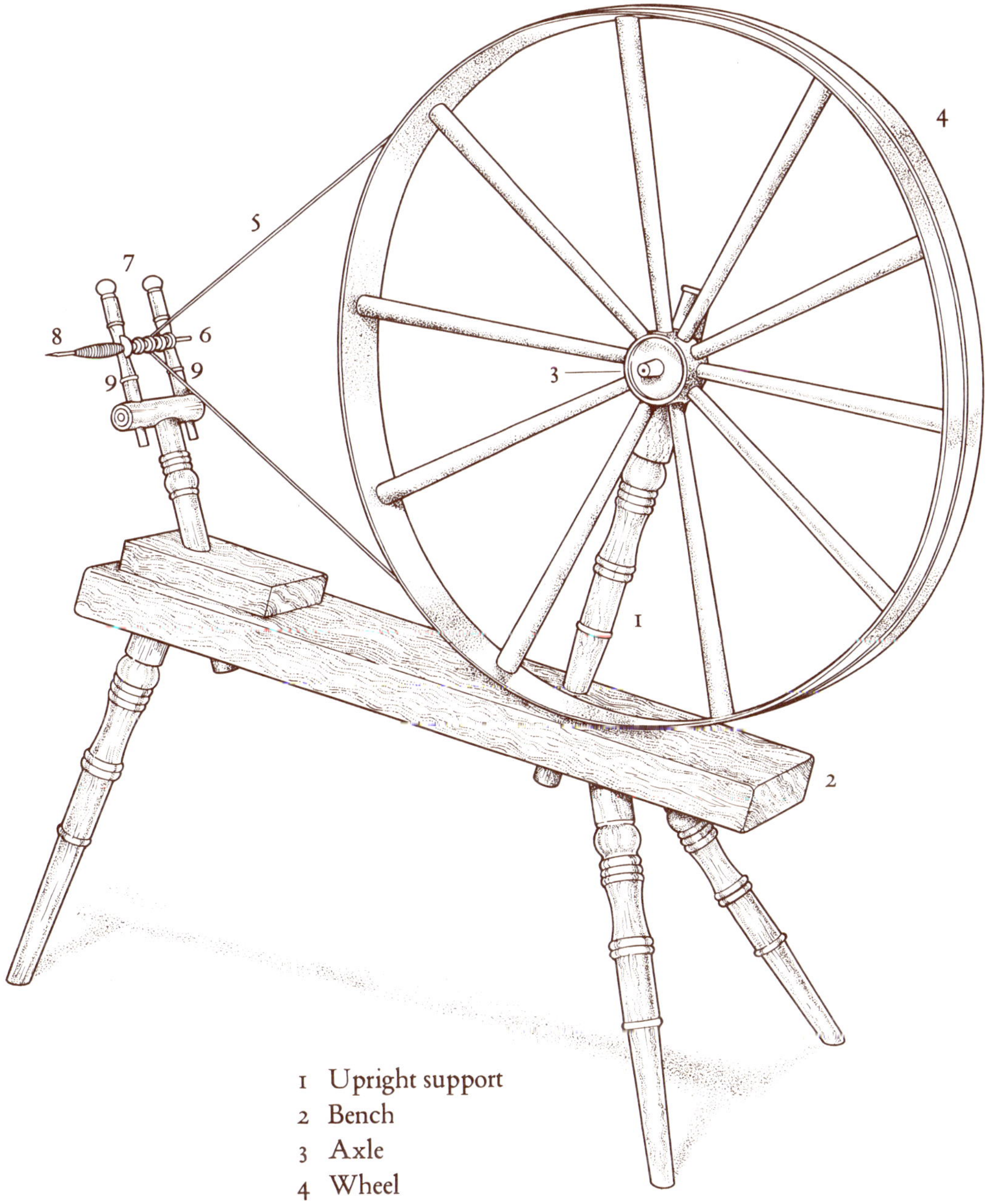

1 Upright support
2 Bench
3 Axle
4 Wheel
5 Driving band
6 Spindle pulley
7 Head
8 Spindle
9 Maidens

1·5 m (60 in). However, 1 m (36 in) was a common diameter for the great wheels used in Europe though the early ones used in households were usually much smaller.

When the first woollen textile manufacturing factories came into being early in the 17th century, mainly in the West Country and East Anglia, the women stood and worked at wheels in rows. The wheels were nearly as tall as they were. In such a case, the distance of the hub of the driving wheel from that of the spindle was of the order of 1·5 m (5 ft) so that, as the spinner or spinster turned the wheel, she had a certain distance to travel from the spindle at the start of spinning a fresh length of yarn to the furthest point at which she could still turn the wheel with her right hand; hence its description – the walking wheel. It has been estimated that, in the course of a day, such a spinster might have covered a distance of some 42 km (30 miles), walking backwards and forwards to her spindle.

One further development of the great wheel which can still be described as such since it did not substantially alter the basic principle on which it worked, was that of the Minor's head. This was the creation in 1810 of an American by the name of Amos Minor. It was simply the addition of a further pulley or gear to work in conjunction with that which was already attached to the spindle shaft (namely the pulley/whorl) and merely had the effect of accelerating the speed of the spindle even further.

Operating the Great Wheel

One thing the invention of the great wheel did not alter was the fact that spinning yarn and winding it on to a shaft or spindle still consisted of two separate actions. It had always been so with the spindle and it remained so with the great wheel. However, since the spindle shaft itself was no longer in a vertical position, either standing on the ground or suspended from a length of yarn, perhaps the way in which it was made to impart twist to the fibres was its greatest triumph.

As the spindle turned, the fibre supply, with the fibres being spun running from it to the spindle, was held in the left hand in such a way that they did not simply wind themselves on to the spindle untwisted and unspun, but slipped off its end. Indeed, only when the spinner had drawn out and imparted twist to a length which took her physically to a point beyond which she could go no further without losing contact with the driving wheel, did she cease the actual

Spinning on the great wheel and reeling the spun yarn; from the Universal Magazine, 1749

spinning part of the operation and wind the yarn she had spun on to the spindle. It has been suggested that, for the purposes of this action, the rolags were made to a precise size so that she would completely exhaust her fibre supply each time she completed the first part of the operation.

To wind the new yarn on to the spindle, the spinner simply altered the angle at which she held the yarn to the spindle from about 135° to 90°, allowing the spun yarn to be drawn on to the spindle evenly to form what is called a cop, by resuming the clockwise rotation of the driving wheel.

So, to start spinning on the great wheel, you should proceed in much the same way as with the spindle: tie on a length of already spun yarn and rotate the wheel clockwise until you have wound a sufficient length on to the spindle to ensure that it has taken a good grip on it. Then stopping the rotation of the wheel with a few inches of already spun yarn still free of the spindle, overlap with these a similar length of unspun fibres which you have drawn out from your rolag.

Perhaps it is worth mentioning at this stage that the major difficulty you will have to overcome in spinning with the great wheel which

Spinning on the Great Wheel

1. The right hand rotates the wheel in a clockwise direction while the left hand draws out the fibres. The twist then creeps up the unspun fibres.

2. Move back as far as possible without the right hand losing contact with the wheel to produce a long length of spun yarn

3. Move in towards the wheel so the spun yarn is held at an angle of 90° to the spindle. Continue turning the wheel in a clockwise direction so the yarn is wound onto the spindle

was not a problem when spinning on the spindle, is that you have to draft the wool from your fibre supply and control the extent to which the twist creeps up the fibres entirely with your left hand. Your right hand is now fully occupied maintaining the rotation of the driving wheel at constant revolutions. So, when you have overlapped the first unspun fibres with the yarn which has already been spun, you must start to turn the driving wheel in a clockwise direction again. In doing so, do not allow either spun or unspun fibres to wind on to the spindle but, holding them at an angle of 135° to the spindle, permit them to slip off its tip. In this way, you will impart twist to both and join them together. Moreover, as you proceed, you will feel the rotations of the spindle start to exert a pull on the fibres in your left hand at the point at which you are holding them and at which you are also controlling the extent to which the twist runs up them. It is at this point, too, that you determine the amount of twist that you give to your yarn as well as by regulating the speed at which the driving wheel revolves.

When you have produced as long a length of yarn as you can, moving backwards as far as you can without losing contact with the driving wheel and drawing out the yarn to the full extent of your left arm as you do so, stop the rotation of the driving wheel. It is now time to wind the yarn on to the spindle.

To do this, you should first reverse the rotation of the wheel in an anti-clockwise direction. Do this until a few inches of the yarn spun at the previous operation are unwound. This permits the twist in those few inches of yarn and that of the next few inches to distribute itself evenly.

Then proceed to turn the driving wheel in a clockwise direction again but this time, holding the yarn at an angle of 90° to the spindle in such a way that it winds evenly on to the spindle. Repeat this process until the spindle is full.

One of the problems in doing this is if you hold the fibres at too great an angle to the spindle (more than 135°) you may create too loosely a twisted yarn because it slips off the tip of the spindle too quickly. On the other hand, if you hold them at too shallow an angle, you will impart too much twist to them. Therefore, either turn the wheel more slowly or hold the fibres at a different angle.

Again, if you grasp the fibres you are holding too tightly, you may retard the rotation of the spindle and cause the driving band to slip on the pulley or the driving wheel with the consequence that, once again, you will impart too little twist to the yarn.

Overleaf: Some of the materials available manufactured from pure wool. The differences in weight and texture are quite marked

Worsted

Twill

Lightweight suiting

Gaberdine twill

Worsted tartan

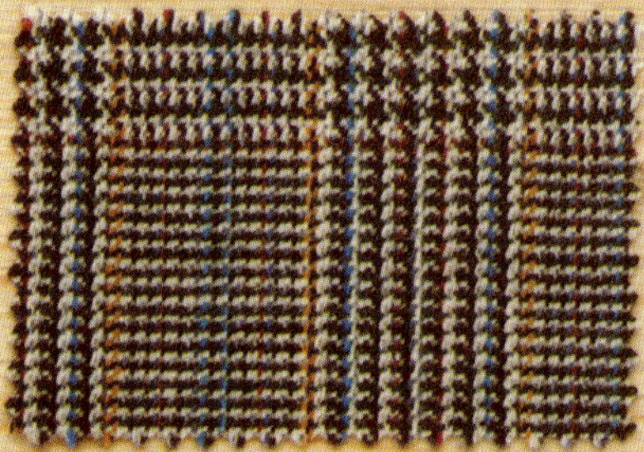

Medium-weight suiting

Wool

Lustrewool

Georgette

Swaledale

Herdwick

Harris tweed

Despite initial resistance to the use of the spindle wheel, it eventually achieved considerable importance. Indeed, it eventually had its place in every household, especially for the spinning of wool, in spite of further developments in spinning technology. For even in the homes of the great and wealthy, wool was still spun, sometimes by the lady of the house, assisted by servants in the combing and carding, and sometimes entirely by the servants because, in time, a certain snobbery affected the spinning of wool which kept it out of the drawing room.

Gentlewomen would therefore only spin flax and do so on wheels which eventually were objects of great beauty and sophistication, especially with the advent of the golden age of English woodworking at the end of the 18th century.

Spinning wool was a means of aiding the family budget. Though it might not be woven into cloth in the same household, it and certainly any surplus would be sold to itinerant wool brokers or wool broggers, either to be made up into cloth by craftsmen weavers for the use of that household or for eventual purchase by third parties. Elsewhere it was made on a more commercial basis. For both during the Middle Ages and at least until the Industrial Revolution, there were families everywhere who depended for a substantial part of their income on spinning.

Hargreaves' spinning jenny

At first these would work at home and have their yarn collected by the wool merchants or clothiers. But subsequently (during the 17th century) these clothiers, presumably having tired of travelling about the country to collect their yarn, organised them to work in factories. Initially this happened mainly in the West Country and in East Anglia. These parts of the country had hitherto been at the forefront of the woollen textile industry in England; particularly East Anglia because it was to here that Flemish weavers fled when persecuted in the 14th century in their native land.

However, it eventually became the general practice, even in Yorkshire. Here, at first those households who had made some part of their living at their spinning wheels, had managed to retain their independence of such clothiers for much longer. This was because they were also farmers who kept their own sheep and therefore produced their own raw materials, instead of having to rely on someone else to purchase them for them. Eventually, however, some of these farmers organised a number of their agricultural workers to spin full-time. Those who organised their workers in this way eventually became the wealthy woollen merchants of Bradford and the surrounding area.

Such a situation was assisted by the profusion of small, fast-running streams from the Pennines. These were employed in the first place to run turbines, in the second place to fill the boilers of steam engines and subsequently to assist in the processing of large quantities of wool and of woollen and worsted cloth. It seems strange at first sight, however, that before spinning was mechanised in the 19th century (the first attempts having been Hargreaves' spinning jenny and Arkwright's spinning frame) it was still the great wheel that was used in such factories.

As we shall see in the next chapter, the spinning wheel had received the attention of highly original minds and had been improved into a much more sophisticated piece of equipment. However, the great wheel persisted.

This can only be because the comparative simplicity of its construction made it cheap and because whatever extra profit, if any, was to be made from investment in more sophisticated wheels, this could not be justified commercially. The handspinners of those days, therefore, were obviously extremely adept at their craft, possibly able to spin yarn just as quickly on the great wheel as on any other kind. No doubt this was the reason why the great wheel remained in general use alongside these other more sophisticated wheels for at least five centuries.

Overleaf left: Hand shearing in Harris, Scotland

Overleaf right: After washing, raw wool is hung out to dry in the sun

MAIZE
FLAKES

Spinning on the Flyer Wheel or Flax Wheel

The Principle Behind the Flyer

The flyer which was invented and attached to the spinning wheel some time during the 16th century, is basically a U-shaped device. Like many other technological improvements and advances, it seems to have been the consequence of the cross-fertilisation of ideas between one sphere of man's activities and another. For like the adaptation of the silk-unwinding machine used in China to the purpose of spinning, it was seemingly an adaptation of something already in use in another industry; namely an S-shaped device used for winding spun silk on to bobbins in the silk industry in Northern Italy.

However, the most important aspect of the flyer's invention is that, for the first time, it made it possible for the spinner to combine the actions of imparting twist to the fibres to create a yarn and that of winding yarn on to a bobbin. For now the yarn was wound not directly on to a spindle, but on to a bobbin which slipped easily on and off it. The yarn passed first through an orifice in the centre of the flyer (also the point at which it was attached to the spindle) then around one of several hooks fixed to either arm of the flyer and thence on to the bobbin.

However, though the diameter of the driving wheel attached to the great wheel came to vary considerably over the centuries, this was not so in the case of the flyer wheel. It seems unlikely that a wheel of any considerable dimensions on this machine would have been practicable. Another major difference between the two types is that the flyer wheel is operated by the spinner from a sitting position and driven by the action of a treadle. Perhaps a driving wheel much larger than about 75 cm (30 in) could not easily be turned by such an action. Yet this, no doubt, also goes to explain why the flyer wheel was generally adopted for domestic use but failed to supplant the great wheel for use in factories.

The flyer wheel was much more compact and its operator could sit comfortably. It was therefore much more suitable for use and storage in the home. On the other hand, the spindle of the great wheel would

turn much faster than that of the flyer wheel, despite its comparative crudeness; thus, what it lacked in mechanical sophistication, it would make up for in sheer speed, particularly in the hands of an accomplished spinster. It also explains why the flyer wheel came to be known, at least in England, as the flax wheel. For this was the type of wheel which ladies employed in their drawing rooms to spin flax; wool, as already mentioned, being banned from genteel society.

As to the invention of the flyer wheel, its accreditation has been and may still be a subject of controversy. Drawings of such a device, or at least similar to it, appear in the *Codex Atlanticus* of Leonardo da Vinci. It has also been attributed to a certain Meister Johann Jurgen of Brunswick in Lower Saxony. However, recently an illustration of a spinning wheel with a flyer has been discovered in a book called the *Mittelalterliches Hausbuch* of about 1480. This is a picture which shows a flyer not in any rudimentary or experimental form but in its final shape and position on the spinning wheel, suggesting that someone had been thinking for some time about the possibility of such an innovation.

It therefore seems more likely that Leonardo's drawings (dated *c.* 1490) represent his experiments with a mechanism which already existed rather than that it was he who had thought of the flyer in the first place. And as to the claim of Meister Jurgen of Brunswick, though it is recorded that it was he who first brought that device to that city, indeed he was given the nickname of 'Spinnrad' or 'spinning wheel', it is also mentioned that the year when he did so was 1530.

The Flyer

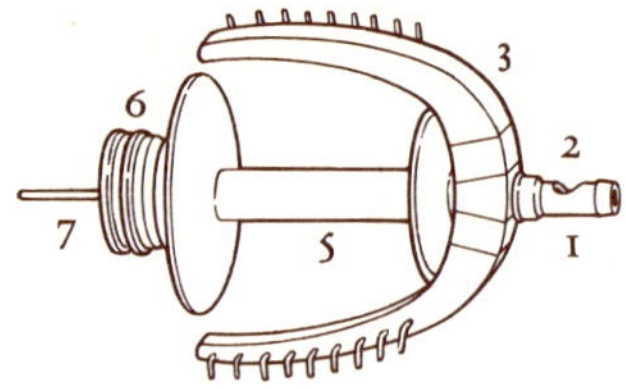

1 Spindle shaft
2 Orifice
3 Flyer
4 Flyer hooks
5 Bobbin
6 Bobbin pulley
7 Drive pulley

Different Types of Flyer Wheel

There are many different kinds of flyer wheel. The two most common types are the sloping table or Saxony wheel and the upright or Shetland wheel. However, there are many others, some of them, such as multiple spinning wheels at which sometimes as many as four spinners could work simultaneously, quite bizarre in their appearance. Then again, there are the castle wheels found in Ireland, which have the flyer mounted below the driving wheel, and modern interpretations such as the Ashford wheel made in New Zealand and a wheel developed in America which is supported on a simple box-like frame.

Like this last mentioned, there are still other wheels that seem to have thrown to the winds not only any sentimental attachment to

Previous page: Hillmen spinning as they tend their sheep in the Himalayas

A copy of the illustration of the flyer wheel in the *Mittelalterliches Hausbuch*

An early 19th-century English table spinning wheel

Overleaf: Spinning on a charka in Pakistan

traditional shapes, but which have also rejected the notion that the power that drives them should be provided by the human body, though they still remain machines for spinning by hand. Such wheels also often comprise of such contemporary devices in their manufacture as bicycle wheels along with sewing and washing machine motors.

However, in most cases, the principle remains the same: that is the principle employed in the two most common types. Indeed, it seems that the major difference between all kinds of spinning wheel really derives from the designer's initial decision where to place the driving wheel in relation to the spindle. For from this emanate subsequent decisions about the overall shape of the structure which supports the mechanism, about the way in which the treadle is mounted and how it transmits power to the driving wheel.

Thereafter, from the basic shapes which had been laid down in this way, further developments such as spinning wheels with two or more flyers and spindles, seem to have been directed by the designer's intention to enable spinners to spin more yarn more quickly.

The sloping table or Saxony wheel

Known as the Saxony wheel presumably because it was for a long time thought to have been invented by Meister Jurgen in Lower Saxony, this wheel has been made and used in most parts of Europe. However, though this may not be exactly an accurate description of it, its other title, the sloping table wheel, is. This derives from the fact that the three-legged stool on which the mechanism is mounted, has an inclined top. It is this which gives this particular type of wheel its unique character, this and the fact that its legs splay out at angles which, it sometimes seems, no designer could ever have determined beforehand.

However, at the lower end of the sloping top or stock is mounted the driving wheel. This usually rests by its axle on two notches cut into the tops of two posts which stand at either side of it. On one end of the axle is a crank which is connected by a length of wood called a footman to the free end of a triangular treadle. This treadle is hinged at its remaining two corners into the feet of two of the legs of the stool. Then the power transmitted to the driving wheel from this treadle is transferred to the spindle by a driving band which, in most cases, runs around the spindle whorl.

However, not all flyer wheels are the same in respect of the way in which the driving wheel drives the flyer mechanism. Some wheels, for example, have a single band which runs from the driving wheel around a groove in one end of the bobbin, not around the whorl on

Sloping Table or Saxony Wheel

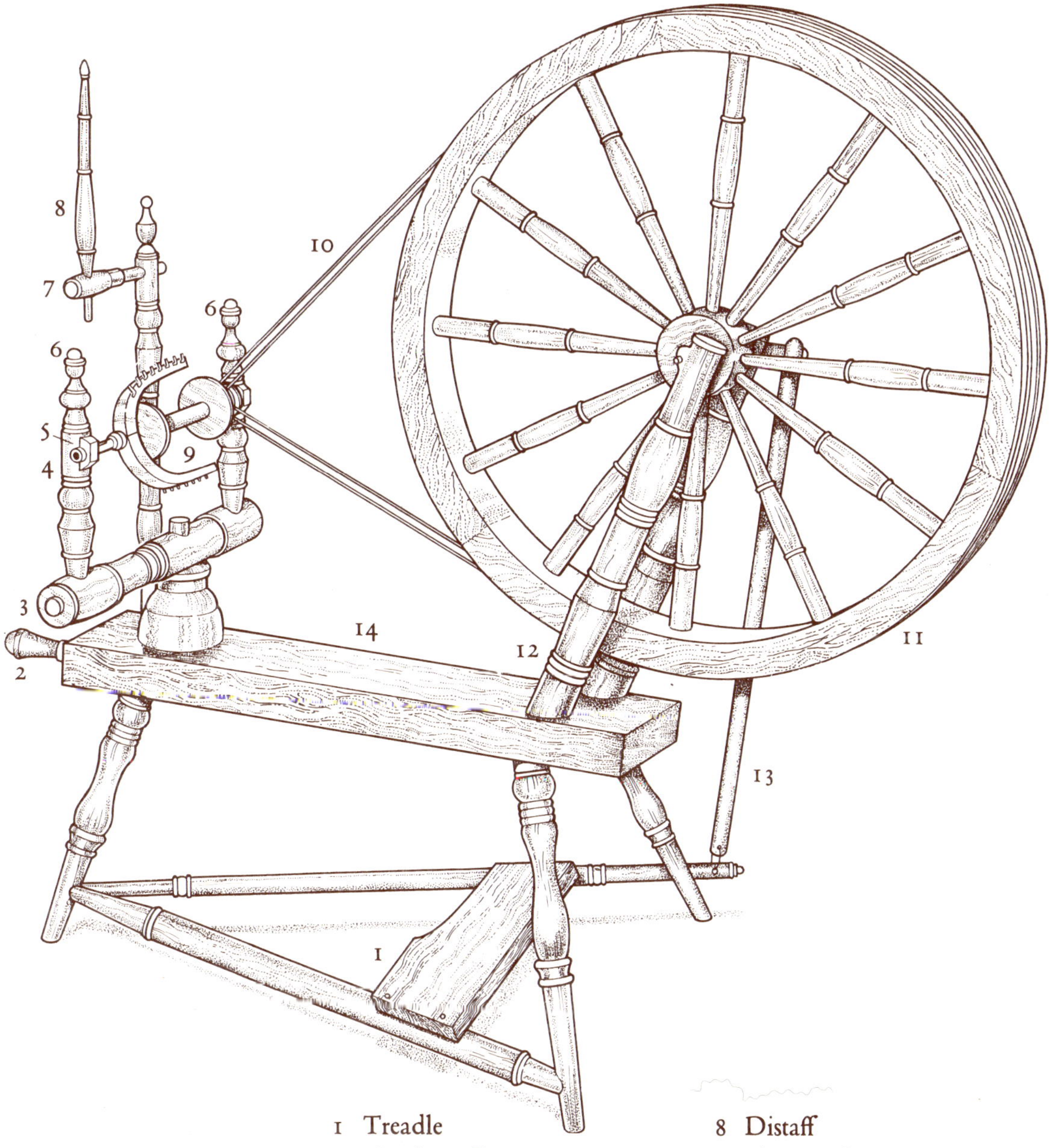

1 Treadle
2 Tension adjustment
3 Mother-of-all
4 Spindle shaft and orifice
5 Leather bearings
6 Maidens
7 Distaff holder
8 Distaff
9 Flyer mechanism
10 Driving bands
11 Wheel
12 Uprights
13 Footman
14 Table

Shetland or Upright Wheel

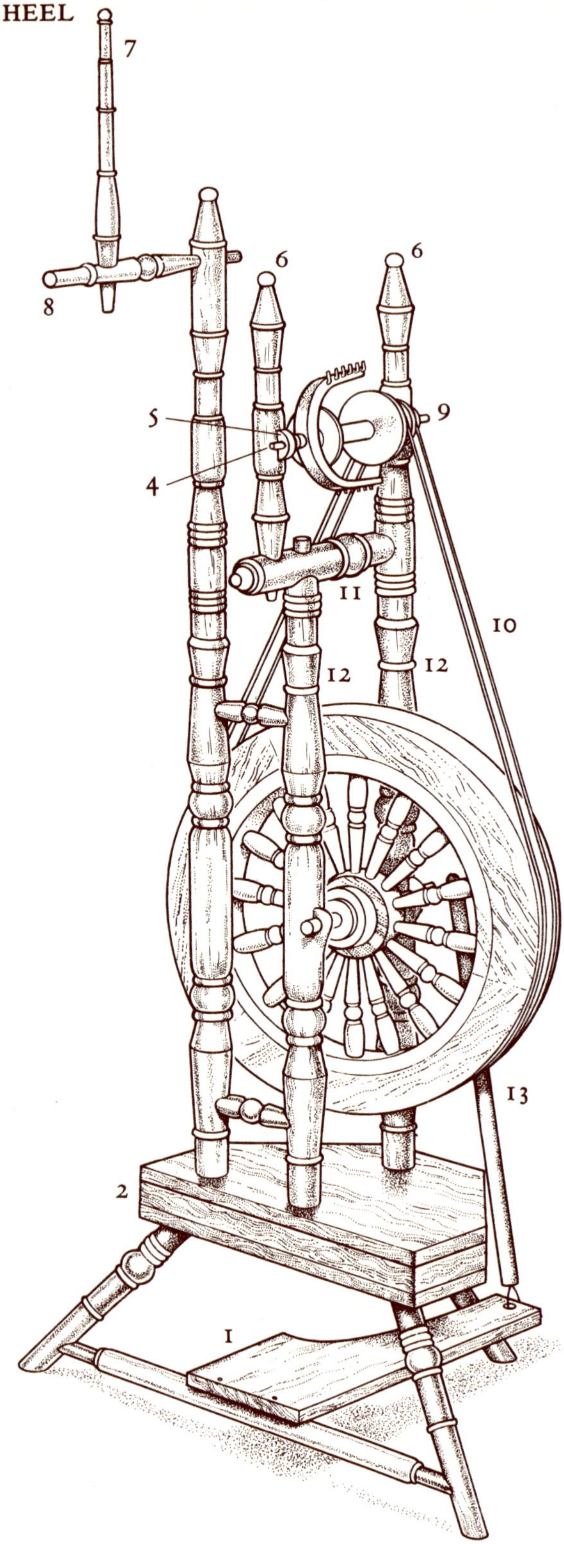

1 Treadle
2 Table
3 Wheel
4 Spindle shaft and orifice
5 Leather bearings
6 Maidens
7 Distaff
8 Distaff holder
9 Flyer mechanism
10 Driving bands
11 Mother-of-all
12 Uprights
13 Footman

the spindle. Others, like the great wheel, have a single band from the driving wheel which runs around the spindle whorl.

Still others, however (perhaps those most commonly found on the flyer wheel), have a doubled band, consisting of a single large loop folded into two. These resulting smaller loops run from the driving wheel; one to encompass the spindle whorl and the other a groove in one end of the bobbin. The mounting of the flyer, spindle and bobbin on the Saxony wheel is between two other smaller posts. These are called maidens and the part of the spinning wheel which comprises of these and the horizontal bar joining them is called the mother-of-all.

It is by moving the mother-of-all either towards or away from the driving wheel that the spinner adjusts the tension of the driving band. Generally this is effected by means of a wooden screw in the end of the stock which penetrates a tenon at the base of the mother-of-all. This tenon slides in a mortice about twice its own length, drawn either backwards or forwards by a turn of the screw.

The Shetland or upright wheel

This differs from the Saxony wheel mainly in that the driving wheel, instead of being mounted to the right-hand side of the mother-of-all is mounted directly beneath it. However, it still rests on two posts which stand on a three-legged stool, though this now resembles a milking stool, having a horizontal top. The posts themselves, too, stand vertically instead of leaning towards the right and, because of the position of the driving wheel, it is also possible for one of them to serve a dual purpose by projecting several inches above the circumference of the wheel to form one of the maidens in the mother-of-all. Meanwhile, the other maiden stands vertically on a horizontal member which connects the top of the other upright post to the taller one. Then the flyer mechanism, incorporating spindle and bobbin, are held as before between the two maidens directly above the axis of the driving wheel where it is allowed to revolve.

As to the tensioning device, this, instead of being a wooden screw set into the stock, is a wooden screw set into the top of the taller of the two wheel uprights. For just as the object of such a screw on the Saxony wheel was to move the spindle whorl either nearer or further away from the driving wheel, so it is with that on the Shetland wheel. However, in order to accomplish this now, it has to be capable of moving the spindle whorl (or the groove in the end of the bobbin) not to left or right but up or down. For it remains true of the Shetland as of the Saxony wheel that it may have either a single or a doubled driving band, driving either the bobbin or the spindle whorl or both.

Every revolution of the U-shaped flyer imparts one twist to the forming yarn. However, one might expect the length of yarn to which it imparts this twist to vary because, as the bobbin fills up with yarn, it tends to draw it in through the orifice at the end of the flyer even more quickly. Yet if this were permitted to occur, it would ultimately lead to unsatisfactory results either when knitting or weaving the resulting yarn. It would mean that the first yarn to be wound on to the bobbin would be more tightly twisted than that which came last, and loosely twisted yarn has a softer look and feel than tightly twisted yarn. Therefore, it is not desirable to mix the two in a single piece of cloth or clothing.

Ideally then, it is necessary for the spinner to control the amount of twist in the yarn and keep it constant. There are in fact several ways of doing this, all of them made possible by the fact that the bobbin is never actually fixed to the spindle but slips loosely on to it.

The first of these is by the simple action of the spinner's hands in retarding the yarn and thereby the revolutions of the bobbin. However, it seems clear that handspinners of the past must have considered this method less than adequate, perhaps because it required too much concentration over a sustained period. Therefore they devised other means either of retarding the revolutions of the bobbin as it filled up or (in the case of wheels which lacked spindle whorls but whose bobbins alone received the drive from the driving wheel) of retarding the flyer at the start of spinning and then allowing it to revolve more freely as the operation proceeded.

It seems, at any rate, to have been precisely this problem which occasioned the adoption both of the three ways in which the driving wheel transmitted its power to the flyer mechanism, and of comparatively crude but effective subsidiary devices to control the amount of twist in yarn either by retarding the flyer or by retarding the bobbin. These are known respectively as flyer drag and bobbin drag and are adjustable as spinning continues. They usually consist simply of a length of cord attached to the wheel between two points in such a way that they bear against either the spindle whorl or the groove in one end of the bobbin, thus creating extra friction.

Like the great wheel, effective spinning on the flyer wheel depends in the first instance on a regular and rhythmic turning of the driving wheel. For irregular or jerky movements of the driving wheel will be reflected in the yarn. Therefore the first thing to consider is achieving a comfortable sitting position at the wheel from which you can operate the treadle smoothly, keeping the speed of the driving wheel constant either in a clockwise or in an anti-clockwise direction.

However, as you do this, you would also be wise to ensure that all parts of the wheel are working properly, are well-oiled and in good order. For obviously, however hard you try, you will never be able to spin to good effect on a wheel that is unserviceable.

If, therefore, the wheel is old, you should ascertain that it has in fact still got all its parts and that they are not badly worn or broken. For example, flyers are notoriously easy to damage because of the short grain across their base, making it easy to knock off their arms. For this reason, it is not unusual to see them bound with string around this part, either to give them extra strength or to help them to resist further cracking.

Sometimes, too, the rows of hooks on the arms of the flyer whose function is to distribute the yarn evenly along the length of the bobbin, are either missing or bent to the point of being unusable. Similarly, a spindle whorl may be missing either completely or in part. In the latter case, of course, even the fact that it lacks just a small segment of its circumference renders it just as useless as if it had not been there at all. The same applies to the bobbin. It too may lack a piece from its end with the groove in it, with exactly the same consequence. As to the driving wheel itself, again if the wheel is old, its rotation over the years may have worn its bearings. Therefore, it will not spin around its centre without wobbling and catching on the posts that carry it.

The only way to discover such things is to attempt to use the wheel and see what happens. However, in theory at least, if you purchase a new wheel, all you should have to do before attempting to start to spin is familiarise yourself with its workings.

The short-draw technique

Of the two different ways of spinning on the flyer wheel, the short-draw technique is probably the most useful and the easiest. It can be employed for spinning either woollen or worsted yarns because it is suitable for use with wool fibres of either long or short staple.

The Short-draw Technique

1. The rolag is held in the left hand while the right hand grips the fibres to prevent the twist moving up the unspun draughted fibres

2. The right hand then moves toward the spindle shaft to draw out the fibres

3. The fingers of the right hand are then relaxed to allow the twist to move up into the draughted fibres

The first step in spinning in either way, however, is to attach some already spun yarn to your spinning wheel. To do this, take about 45 cm (18 in) of yarn and pass it through the orifice in the end of the flyer. Then, having taken it over one of the rows of hooks on either arm of the flyer (it does not matter which) lead it down to the axle of the bobbin. Then run it round this two or three times and tie it on tightly.

When you have done this, the spinning by this method on the flyer wheel resembles, in most respects, spinning on the suspended spindle. For except for the fact that you are now spinning with the fibres being drawn out and into the orifice of the flyer horizontally, your hands positioned accordingly, the rest of the operation in principle remains the same.

In the first place, you join the new fibres on to the already spun yarn in the same way, overlapping them, turning the wheel and allowing them to twist together. In the second place, the left hand (if you are right-handed) controls the fibre supply, spreading out the fibres to create a triangular drafting area, its base between the thumb and forefinger of the left hand and its apex at the point where the right hand controls the twist. And again, it is the length of the fibres being spun that determines the extent of the drafting area.

For if you attempt to impose a drafting area which is too short on your fibres, you will find yourself gripping them at both ends and not permitting them to slide past each other. On the other hand, if you hold your hands too far apart, you will find it difficult to control the tension in the drafting fibres, with the result that they will not draft evenly.

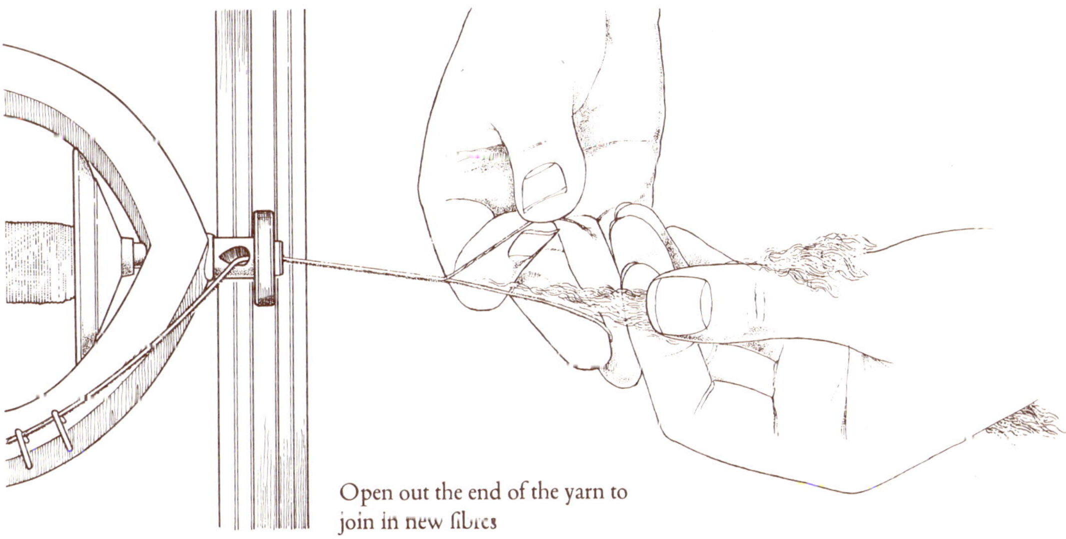

Open out the end of the yarn to join in new fibres

1. The rolag is held in the right hand while the left hand grips the fibres

2. Draw out the fibres with the right hand until the right arm is fully extended

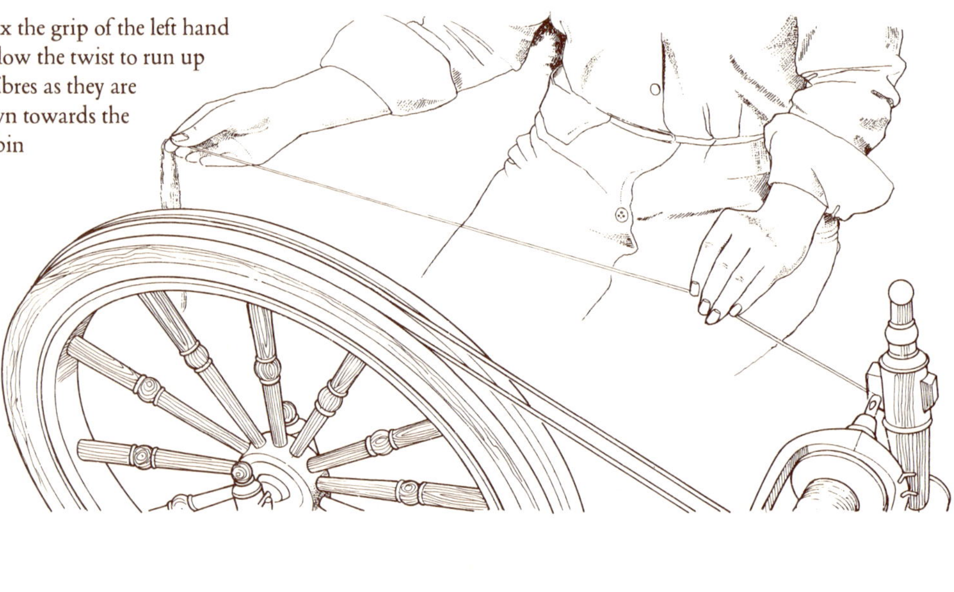

3. Relax the grip of the left hand to allow the twist to run up the fibres as they are drawn towards the bobbin

The long-draw technique

The main difference between this and the short-draw method is that the left hand does not permit the wheel to impart twist immediately to the fibres being drafted. Rather it prevents much of this twist running up the fibres until the right hand has drawn them out to a considerable distance. Moreover, in gripping these fibres and holding them, it also retards the rotation of the bobbin, preventing any yarn being wound on and causing a considerable degree of twist to build up between itself and the orifice.

Indeed, it is only when the right arm is fully extended away from the wheel that the left hand is allowed to release its grip, thus permitting both the twist to move up the drafted fibres and the newly formed yarn to be wound on to the bobbin. So, in practice, to the onlooker, the two operations seem distinctly different.

In contrast, spinning by the short-draw method presents the appearance of one steady, continuous action, consisting of drafting the fibres between two hands, held constantly in the same position, and simultaneously allowing them to be twisted and wound on. Spinning by this long-draw method, on the other hand, represents two rhythmically executed but separate movements, the left hand held only a few inches away from the orifice, the right hand moving over and over again, first away from it then towards it. Even so, initially, the method is the same as for the short-draw technique.

Attach a length of already spun yarn to the bobbin, having already passed it through the orifice and across the hooks of the flyer. Join the new fibres on to the spun yarn as previously described. It is now that the instructions start to diverge.

Beginning with your left hand only a few inches away from the orifice and your hands close together, grip the fibres tightly in your left hand as you treadle, allowing twist to build up in front of it. Meanwhile, draw your right hand back, drafting out the fibres in the fibre supply until you have drawn them out to the fullest extent of your right arm. Then release the grip of your left hand on the yarn and allow the twist to run up the drafted fibres as they are drawn on to the bobbin.

Obviously you must allow some twist through the fingers of your left hand while you are drafting, otherwise the fibres you are drafting will disintegrate. However, a gentle twist in the fibres as you draft them will make this not only possible but easier.

Preparing the Yarn

Removing Your Yarn From the Spindle

What you have when you have spun a full spindle, cop or bobbin is an indeterminate length of single-ply yarn. This may be plain yarn, of woollen, semi-worsted or worsted, or a textured yarn of varying thickness. Whichever it is, it may be exactly what you require if you are proposing to make something delicate like a crocheted shawl. On the other hand, it will probably be too thin for knitting wool. Therefore, before you go any further, you will want to take the yarns from two or more spindles, cops or bobbins and ply them. However, whatever kind of yarn you want, the methods you eventually employ to remove it from these will vary little.

The next stage you are preparing your yarn for is washing or scouring. When you take it from spindle, cop or bobbin, you need to arrange it immediately in an open but still orderly form. Such an arrangement is known as a skein. If you were to remove the yarn from your spinning machine simply by winding it into a tightly-closed, compact ball, this would be unsuitable as when you came to try to wash it, the natural oils would be unable to escape and the suds would be unable to reach them.

There are many ways of achieving a skein. It has been suggested that in primitive times yarn was unwound on to a forked stick or antlers. Indeed, the designs of the devices more recently employed in skeining tend to support this proposition. For though they differ in size and shape, they remain the same in principle.

Such an instrument is called a niddy-noddy. No doubt it is from the nodding motions of its head and tail when in use that it gets its peculiar name. In any event, basically this is just a length of wood with two shorter lengths attached at their centres across both ends. It thus forms the general shape of a letter H. However, it differs from this in that the two end-pieces are set at right angles to one another, their points projecting respectively north-south and east-west.

It is between and around the four outstretched arms of this instrument that you stretch and wind your newly formed yarn in strands. The yarn is run from top to bottom, first around an arm on one side of the shaft and then around an arm on the other until

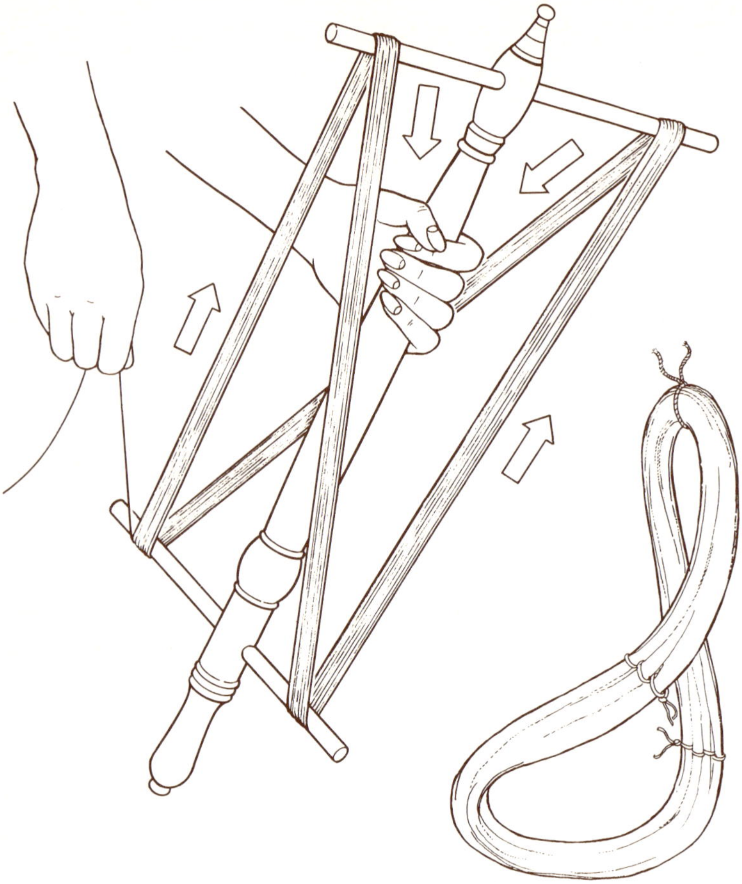

Left: The spun yarn can be wound onto a niddy-noddy to form a skein

Right: The finished skein correctly tied to prevent tangling during washing or dyeing

spindle, cop or bobbin is empty. When this is empty, the yarn must be removed from the niddy-noddy for washing. For this purpose, most niddy-noddies have at least one easily removable arm which will enable you first to release all the loops of yarn which run around this and then those which run around the other arms in natural succession.

When you have done this, you will find that, by holding the loops of yarn you removed first together with those you removed last, you will achieve a skein rather shorter than the distance between the two ends of the niddy-noddy. In this skein, the individual strands twist around each other loosely and take up their length in such a way that the overall appearance of the skein is that of a corkscrew. This is precisely what is wanted because in this form it will be possible both for you to wash it effectively, removing all dirt and oils, to dye it in such a way that the colourant can reach all the fibres and, in either process, to control the individual strands of the yarn so that, when wet, they do not become tangled and unmanageable.

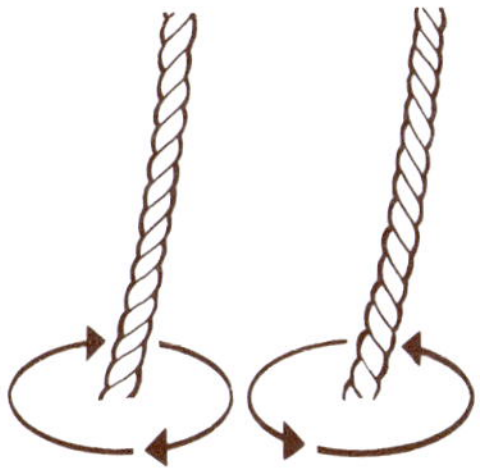

If you do require to ply yarns, there are as many ways of doing so as there are of spinning. Similarly, there are at least as many different effects you can achieve by plying as there are by spinning. For though the yarns most commonly plied for knitting are probably Z-twist, evenly spun and consisting of two yarns only, there is no reason why you should not, if you desire, ply three or more yarns together, some of which may be S-twist and some Z-twist, to create some special effect.

In addition, since yarns may initially be spun thick or thin, plain or slubbed, under or overplied, so you may combine these characteristics in any way you please to multiply the range and variety of yarns you may obtain eventually. You can obtain a spiral yarn when you are plying by gripping one of the yarns being plied more firmly than the others.

However, as to the methods you may employ to ply yarn, you can use any of the machines that you might use for spinning in the first place. Therefore, the principle ways in which you can ply are with the assistance of spindles or of the great wheel or the flyer wheel.

Plying with spindles

To ply with spindles, for instance, using two or more full spindles depending on how many yarns you want to ply, take the ends of the yarn from each and attach them to the empty shaft of a further one. Then suspend this spindle from these yarns in the same way as you would for ordinary spinning. Meanwhile, sling the yarns from the two or more full spindles over your shoulders and allow the spindles themselves to rest on the floor behind you. Then, rotating the suspended spindle (but this time in an anti-clockwise direction if you are plying Z-twist yarns) and imparting the desired amount of twist to them, draft them out between thumbs and fingers of both hands until the spindle touches the floor, stop and wind on the newly plied yarn. Repeat this operation until you have used up all the yarn on the spindles behind you.

However, you can accomplish the same task if you possess only one spindle. You do this by removing the yarn from it every time it is full and winding it either into a ball or on to a stick.

When you have a number of balls or sticks of yarn you wish to ply, then place them either in bowls or on the ground, take their ends and attach them to the spindle as before. Then, with the bowls containing the balls of yarn or the sticks of yarn behind you, run the yarns from them over your shoulders again and suspend your spindle from them

Plying with Spindles

Always twist the spindle in the opposite direction to that used when spinning the single-ply yarns: clockwise when plying S-twist yarns or anti-clockwise when plying Z-twist yarns

in front of you, spinning it in an anti-clockwise direction. Having done this, wind the plied yarn on to the shaft of the spindle as you would when spinning ordinary single-ply yarn.

Plying on the great wheel

Again, you ply yarns on the great wheel in the same way as you spin yarn on the great wheel – in two separate actions, first imparting twist to the lengths of yarn, then winding them on to the spindle. Indeed, the only difference between plying on the great wheel and on any other spinning device is that, instead of turning the driving wheel in an anti-clockwise direction when plying Z-twist yarns, you persist in turning it clockwise.

You make this possible and yet still succeed in reversing the direction in which the spindle turns, by crossing the driving band. To do this, release the tension on the driving band, remove it from the driving wheel and twist the loop so that it forms a figure-of-eight. Then replace it on the driving wheel and proceed to ply, still turning the driving wheel in a clockwise direction.

Plying on the flyer wheel

When plying on the flyer wheel, however, you do reverse the direction in which the driving wheel turns. Moreover, when plying on this device, you do not have quite as much scope as regards the thickness of the yarn you can make as you do on other machines. For the size of the orifice at the end of the flyer on your wheel naturally limits the diameter of the yarn that will pass through it. Thus, you will not be able to ply any number of yarns together as you could on the great wheel or on spindles, but only as many as can pass through this aperture together comfortably. However, this does not mean that plying thick yarn with the assistance of the flyer wheel is impossible.

You can adapt the flyer by plugging a length of dowel of suitable diameter into the orifice on the flyer and using this as the spindle with which you ply your yarn, and then proceeding in the same way as you would ply on the great wheel. Although the flyer is inoperative using this method you still benefit from the action of the treadle.

However, particularly in regard to the flyer wheel, there does exist a subsidiary device to assist you when plying. This acts by holding the bobbins you are drawing the yarns from and the device is called the Lazy Kate. It holds the bobbins side by side on a long spindle mounted in a box-like frame, allowing them to revolve freely and unwind the yarns evenly as they are drawn off by the spinning wheel.

Plying on the flyer wheel; the bobbins are held on a Lazy Kate standing on the floor

Otherwise, the only aspect you have to consider when plying on the flyer wheel is to keep the amount of the twist in the plied yarn constant. As when spinning on it, control the twist either manually by retarding the progress of the yarn with your fingers when appropriate, or where it applies, by applying drags to the flyer or bobbin.

Washing and Scouring

As already stated, you can choose whether to wash and scour your wool before spinning or after. However, unless your fleece is particularly dirty or unless you want, for some reason, to dye it in its unspun state, then there seems little point in washing it first because the natural oils in the fleece assist you in the spinning and scouring will remove these. On the other hand you may wish to obtain a composite colour of wool by dyeing two or more different batches of it different colours and then blending them together at the carding stage, then you will need to ensure that all the natural oils are removed from the fibres, otherwise they will not accept the colourant.

By contrast, even if your fleece is dirty, there is no reason why you should wash it so thoroughly that you remove the natural oils. You can cleanse it sufficiently simply by soaking it in warm water without soap because wool contains its own natural detergent. This will remove the dirt without removing the natural oils.

However, whether you are washing it for this purpose or in preparation for dyeing, you should handle your wool carefully. Never wring it to remove excess liquid or subject it to sudden changes of temperature as you might if, for example, you were to place it in running water. Woollen fibres are at their most vulnerable when wet and such treatment tends to cause them damage with the consequence that they matt or felt.

Therefore, if you want to clean a quantity of fleece, ready for spinning, first immerse it gently in tepid, soapless water. This will turn dark fairly quickly. You should then gently remove the wool and get rid of the dirty water. Repeat this process until you judge that you have removed most of the dirt without taking the grease out of it. You should be able to detect how much grease remains in it by lifting some out of the water and feeling it with your fingers.

To dry it, you should hang it outside, perhaps on a wall, fence or tree, turning it occasionally to allow the air to get to it. However, in order to remove from it the weight of water that remains in it directly

Lightly spray the fleece with a mixture of olive oil and water if it has been washed and dyed before being spun

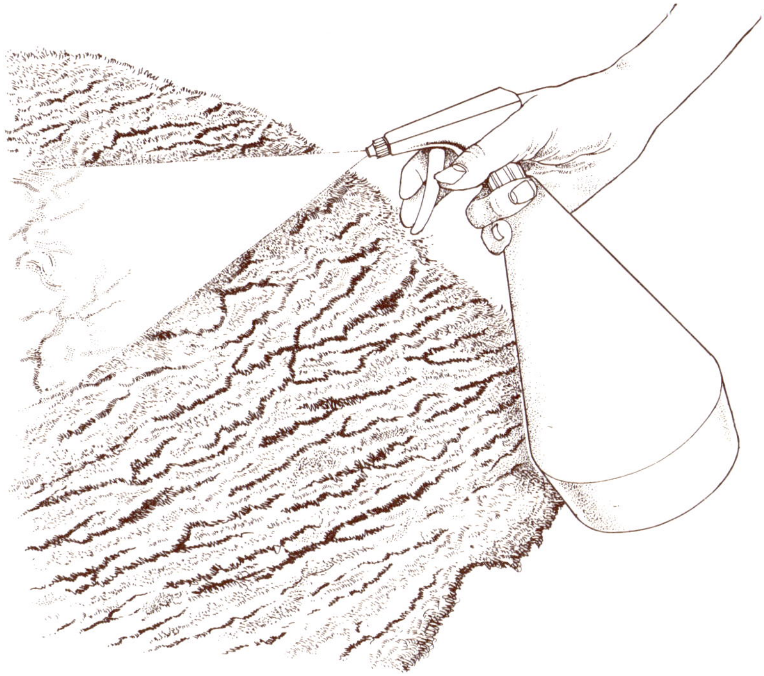

after you have taken it from the tub, you will probably find that you can afford to give it a short spin in a washing machine or spin drier. Such treatment apparently does not amount to rough handling.

The same applies when you have thoroughly cleansed a fleece in preparation for dyeing. However, to do this after the initial soakings, you should wash the fibres first in tepid, then warm, then warmer water with the addition of soap, until you have washed it in hot water, rinsing it out each time in clean water of a similar temperature. Remember that in between each wash and rinse, you should not wring but squeeze or press the excess liquid from your fleece with your hands, gently.

As to the use of detergents, some spinners do employ these for scouring. However, ordinary soap is adequate for the purpose and, as detergents may damage the fibres of your wool, soap alone is recommended. When you have scoured the fleece, dried it and subsequently dyed it as you require, then oil it. Olive oil, or any natural vegetable oil is suitable. If you mix it with water (in proportions of two to one) you will probably find that the most effective way to apply it to your wool is with a spray. However, there are two points to note about this. First, do not oil more fleece than you can spin up within a week to ten days of oiling. Secondly, wash the oil from your yarn as soon after spinning as you can. Some oils, if they are left in the wool, go rancid.

Dyeing

Dyes – Natural or Synthetic?

If you want to dye your wool either before spinning or after, you can if you wish resort to the range of chemical dyes available from your local chemist or hardware store. These are cheap, easy to use and reliable. However, despite such obvious merits, it may be that they do not possess the one quality that we are seeking when we dye handspun wool – the quality of naturalness. What this means is only to be learned by comparison of the colours obtained both with chemical dyes and with dyestuffs obtained from natural sources.

These are materials our ancestors used when they dyed their handspun wool. The quality of colours they were able to obtain from them is still discernible in the textiles made by primitive peoples today and in antique oriental carpets. Moreover, since the first chemical dye was invented in 1856 and since handspinning itself represents a return to the old ways of doing things, dyeing handspun wool with natural rather than chemical dyes somehow seems more appropriate. Yet plainly, if we do choose to use natural dyes, we have to suffer certain disadvantages.

In the first place, the materials from which they are made are relatively difficult to collect. Therefore either we have to pay more for them than for chemical dyes or else spend more of our own time collecting them. Secondly, the recipes for use with them are fairly complicated and again take up a lot of time both in organisation and execution. Finally, it is by no means predictable exactly what shade of a colour will result from a particular recipe. Such results depend not only on method but also on such things as, for example, at what time of the year the dyestuff was harvested. They are therefore subject to whatever variations are to be found in nature, such as different climatic conditions from one region of a country to another, and cannot be tied down to any precise chemical formula.

DYES – ADJECTIVE OR SUBSTANTIVE?

Generally, natural dyes are divided into two categories – adjective and substantive. The difference between them is that substantive dyes are sufficient in themselves to produce a fast colour whereas adjective dyes have to be used in conjunction with some other substance, called a mordant.

Perhaps the most important group of substantive dyes are produced by lichens. These, however, should be used and collected conservatively. As they grow extremely slowly, about 1 mm per year, and as it is generally true of natural dyestuffs that you need approximately 0·5 kg (1 lb) of dyestuff to colour 0·5 kg (1 lb) of wool, plainly a cavalier attitude to their collection and use would be destructive.

Of the mordants used with all other types of dyestuff, however, tin, alum, chrome and iron are the most successful. Alum is particularly suitable for mordanting wool though if you use too much it may cause the wool to become sticky. Tin, in the form of crystals of tin, muriate of tin or stannous chloride, is good for producing bright shades of red and yellow and indeed, may be used both as a mordant and as an additive to dye for brightening colours. However, if you use too much, you may cause your wool to become hard or brittle. Chrome, on the other hand, in the form of bichromate of potash, makes wool soft and silky. However, it is sensitive to light and both the mordant itself and wool mordanted with it should be stored in a dark place until required. Besides this substance seems difficult to obtain, not being generally available from chemists, so that it may be that the only way you can obtain it is either to buy it in a very large quantity (25 kilos (56 lb)) from a chemist or to persuade a tannery to let you have some since it is also used in the preparation of leather.

As for iron, this tends to dull the colours. It is therefore nearly always used in conjunction with cream of tartar which is often used with other mordants for the purpose of brightening. In addition, the method employed for mordanting with iron may be exactly the reverse of that used with other mordants. Instead of boiling the wool in the mordant first, removing the wool, adding the dye and then reintroducing the wool to the liquid, sometimes it is first boiled in the dye. It is then mordanted by the addition of the mordant to the dye, though in such cases the wool should again be removed from the dye during the addition of the mordant. However, as with all the mordants mentioned above, the mordant itself should first be dissolved in water before being added to the dyebath.

As to the quantities of mordant used per pound of wool, these are as follows:

75–110 g (3–4 oz) alum, depending on whether the wool is fine or coarse

10 g ($\frac{1}{2}$ oz) tin crystals

5–10 g ($\frac{1}{4}$–$\frac{1}{2}$ oz) bichromate of potash

10 g ($\frac{1}{2}$ oz) ferrous sulphate (copperas)

1 oz of cream of tartar will suffice to brighten colours mordanted with alum and copperas and 2 oz for those mordanted with tin.

As to the method employed:

(1) dissolve the mordant in water,

(2) add it to the dyebath filled with water,

(3) allow the water to become warm on a stove,

(4) immerse the wool and gradually bring the solution to the boil. However, do not allow it to boil continuously but let it simmer for $\frac{3}{4}$–1 hour, depending on the thickness of the yarn or fibres being mordanted

(5) lift the wool gently out of the dyebath, let it drain, squeeze it gently or press it to remove excess liquid

The wool may now be dyed immediately but is better left for a day or two to recover its vitality.

In some cases, you may add your dyestuff to the water containing the mordant. However, you will get brighter colours if you execute these operations separately.

Finally, note that, before immersing a skein of wool in either mordant or dye, tie the strands together with a length of yarn, loosely but firmly – if you tie them too tightly, you may prevent the dye reaching some of the fibres. If you do not tie them at all, they will almost certainly become entangled.

Lifting the skein of wool out of the dye bath with a rod to check the intensity of the colour

Dyestuffs – Homegrown or Imported?

The dyestuffs available to you to some extent depend on the part of the world in which you live. For example, many are to be found amongst the local flora and perhaps even fauna and the composition of these obviously varies from country to country. However, this is only part of the story. For though dyers were exclusively dependent on dyestuffs to be found locally until the 16th and 17th centuries, subsequently new dyestuffs were imported from abroad as new continents were opened up and their raw materials exploited. Naturally, in many cases then, dyestuffs from abroad superseded those obtained from local sources.

However, at least one group of those already in use in Britain remain in favour. These are the lichens. Perhaps the reason for their continued popularity is that they are substantive dyes and require no mordanting. At any rate, they are responsible for many of the colours to be found in tartans.

But wherever you live and whatever the source from which you choose to derive your dyestuffs, these are likely to be of three kinds:

(1) They may be of animal origins such as cochineal which is made from the small, crushed bodies of South American beetles

(2) They may be derived from plants such as blackberry, heather and bracken and other vegetable sources such as tree bark

(3) They may be derived from a mineral source such as iron rust This produces a deep red and is responsible for the colour of the sails of fishing boats in the Mediterranean.

Their collection, particularly if they are from vegetable sources, should be accomplished while the plants are still young and vital. Lichens are best collected either in winter or after long periods of rain because they can then be removed from the bark of a tree or from stones more easily. However, it is worth noting that the lichens which yield the best dyes are those which have grown on stones.

Then, when you have collected them, they should be stored in a dry place, otherwise they will soon go mouldy. Alternatively they can be dried but generally fresh dye materials give a more vivid colour.

Methods of Dyeing

Basically there are three methods of dyeing. First you mordant your wool and then dye it in a separate dyebath. Alternatively, you mordant and dye the wool in the same dyebath. This is particularly suitable for reds and yellows. Finally, as with iron mordant, you dye your wool first and then fix the colour by mordanting. This process is called saddening as it tends to dull the colours. However, it also makes them very fast and is particularly favoured for use with Kutch and Logwood dyes.

As to the other details, if you are going to dye your wool immediately after mordanting, remember that the temperature of your wool is already near the boiling point of water. Therefore, put your dyestuff into cold water and then bring it slowly to the boil, making sure that it has released as much of its colour into the liquid as possible before you immerse your wool. Otherwise, if you choose to leave your mordanted wool for a few days before dyeing, you should ensure that your wool is at the same temperature as the dye before immersing it. You should do this in one of three ways: either by allowing the dye to cool until lukewarm before immersing it; or by bringing the temperature of your wool up to the boil in a separate pot of plain water before transferring it to the dyebath; or, as is sometimes done with lichens but is not possible with all kinds of dyestuffs, by putting both wool and dyestuff into a pot of cold water and bringing them to the boil together. Furthermore, if your dyestuff is comprised of material which will not entirely dissolve but which may get entangled in the wool you are dyeing, it is a good idea either to place it in a muslin bag in the dyebath or to separate it from the liquid in the dyebath by straining before immersing your wool.

Do not stir your wool too much or stir it with sticks that are badly stained from previous dyeings. If you turn all your wool over once or twice during the course of dyeing, this should ensure that the dye penetrates to all the fibres to produce an even colour. Similarly, if you use clean sticks for turning, this will ensure that your wool is not stained by the colour of some unwanted dyestuff that can easily become ingrained in wood.

Finally allow your wool to remain in the dye until it reaches the desired colour. In doing so, remember that wet wool looks darker than dry wool so that you should aim for a slightly darker colour of wool than the one you want while it remains in the dyebath. Furthermore, it is worth mentioning that you can use the same dyebath for colouring a further batch or batches of wool if you want to obtain progressively lighter shades of the same colour. For each successive dipping in the dyebath removes from it some of its power to colour.

After dyeing, wash your wool thoroughly in hot, soapy water, then rinse it in clear water as many times as you have to until colour no longer escapes from it. If you then wash the wool again in hot, soapy water, this may serve to brighten and fix the colours especially of reds and yellows.

Utensils Needed for Dyeing

Though you will not need the same utensils for dyeing by every recipe, the following list represents most of those that you are likely to need for the purposes of dyeing. The first point to make is that whatever the tool, you should keep it specially for dyeing as this will subsequently render it unsuitable for any other purpose.

The first necessity is a dyebath. This should be of stainless steel or unchipped enamel because dyeing involves chemical reactions which would be affected by the use of, say, a brass, copper or iron vessel. However, galvanised iron is adequate as long as, when you mordant with tin, for example, you make sure that you dissolve the crystals of tin completely in water before putting them in the dyebath. Otherwise they will attack the protective coating. Apart from this, it should also be large enough to hold the quantities of wool you will be using easily and without requiring that you compress them, otherwise the dye will not reach all the fibres.

Moreover, since dyeing often involves more than one operation, it is a good idea to have a separate dyebath for each of these processes as well as smaller ones for dyeing samples. The latter, however, need not be large and may therefore be of heatproof glass since this will not interfere with the intended chemical reaction. In addition, you will need a special dyebath for use only when dyeing with indigo.

As to the other utensils, secondly you will need scales for weighing out quantities of dyestuffs and chemicals. For this purpose, kitchen scales are adequate although you may need something more sensitive for measuring smaller quantities.

Thirdly, a thermometer, although this is really only necessary when dyeing yellows and indigo where the temperature of the dyes are critical.

Fourthly, non-metallic or stainless steel spoons are also needed for measuring chemicals.

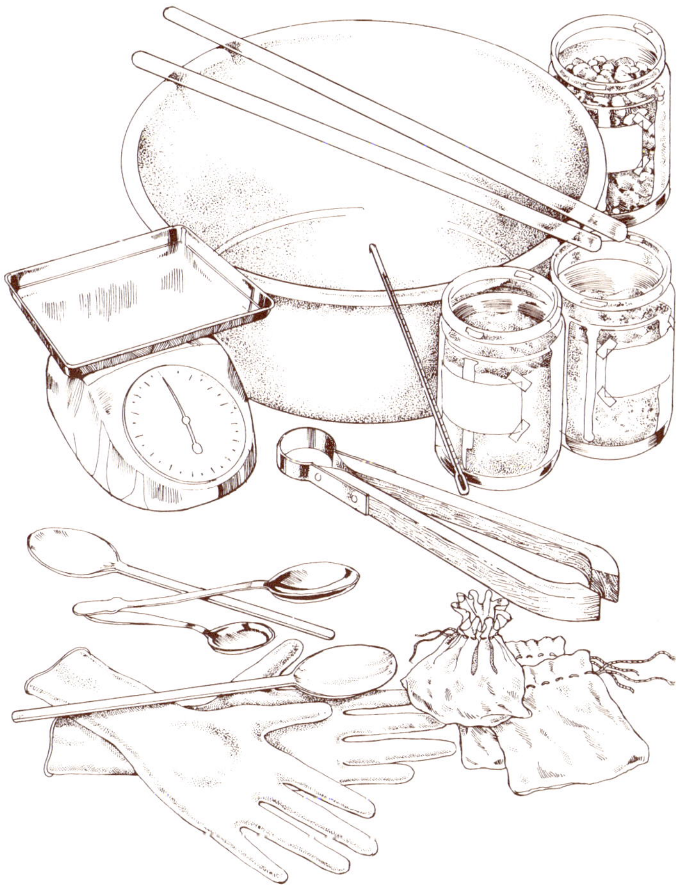

Utensils for Dyeing

Dye bath, scales, thermometer, spoons, stirring rods, rubber gloves, muslin bags, tongs and labelled glass containers for dyestuffs

Fifthly, stirring rods – if these are of wood, you will need to renew them as they become stained because they may stain your wool. Glass ones on the other hand will not do this, though they are more expensive.

Sixthly soft water – hard water spoils woollen fibres so that, if you have no other source, you should collect sufficient quantities of rain water.

Finally, a number of accessories such as rubber gloves to protect your hands from staining, muslin bags to keep your dyestuffs separate from your wool and tongs for lifting it out of the dyebath, although you may accomplish this nearly as easily with a stick or a wooden spoon. In addition, you should have at the ready labelled containers for your dyestuffs and keep these in a particular place out of reach of children as some of them are poisonous.

As you will see from the following catalogue of recipes, there are many different ways of obtaining more or less the same colours. Which one you choose will therefore depend most on what dyestuff is most easily available to you and which your experience eventually tells you produces the colour you want most successfully. Indeed, in this regard, experience in dyeing is plainly of special importance since no mere words or adjectives suffice to describe the exact shade you will be able to obtain with any particular dyestuff or recipe. However, whatever colours you find you can achieve with these, it is equally clear that you can increase their range by dyeing your wool first in one colour, then in another. In this way, you can obtain olive green, for example, after mordanting it with alum, dyeing it first yellow with onion skins and then, having thoroughly rinsed it, with indigo.

Dyestuff	*Part of plant used*	*Source*	*Mordant*
ORANGES			
Coreopsis	flowers	f	T
Golden Rod	flowers	f	T
Onion	whole	f	T
Sloe	bark	f	A
Weld	whole plant	f or d	T

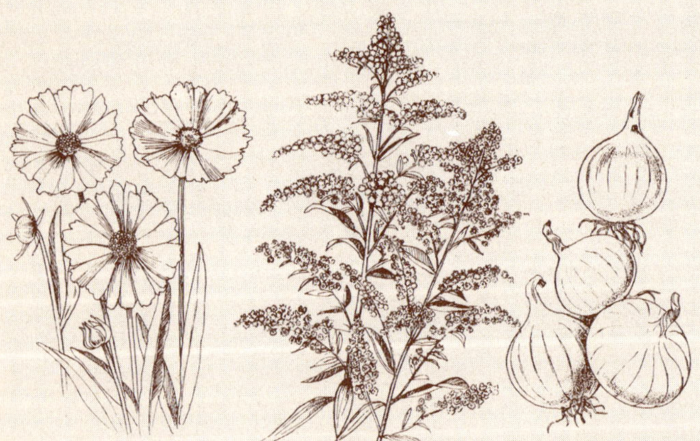

Coreopsis Golden Rod Onion

KEY

- f collect and use while fresh
- c available commercially from chemists or craft shops
- d may be collected fresh and used dried
- A mordant with alum
- C mordant with chrome
- T mordant with tin
- I mordant with iron

Dyestuff	*Part of plant used*	*Source*	*Mordant*
YELLOWS			
Agrimony	leaves and stalks	f	A or A
Bedstraw	tops	f	A or C
Birch	leaves	f	A
Broom	flowering branches	f	A or C
Coreopsis	flowers	f	None
Dahlia	flowers	f	A or C
Dog's mercury	whole	f	A
Fumitory	flowers	f	A
Fustic	bark	c	A or C
Golden marguerite	flowers	f	A or C
Gorse	flowers	f	A or T
Juniper	berries	f	A
Mullein	stalks and leaves	f	A or C
Privet	leaves	f	A
Queen Anne's Lace	flowers, leaves, stems	f	A
Ragwort	flowering tops	f	A
Sea Holly	whole	f	A
Tansey	flowering heads	f	A
Teasel	leaves, flowers	f	A
Viburnum tinus	leaves	f	A
Wayfaring tree	leaves	f	A
Weld	whole plant	f or d	A or C

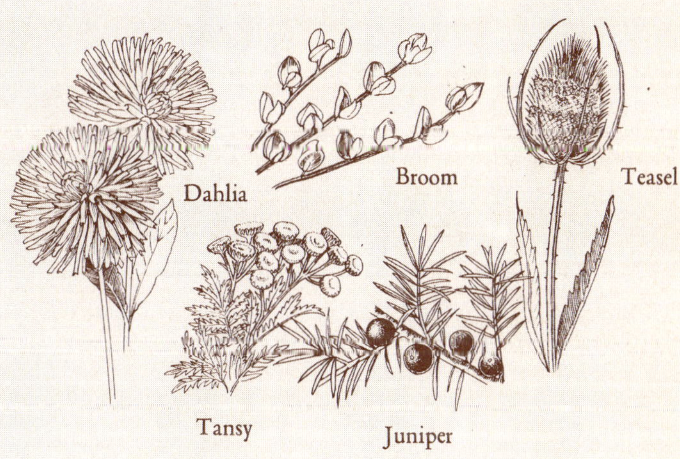

Overleaf: Samples of natural dyed wool

Cochineal

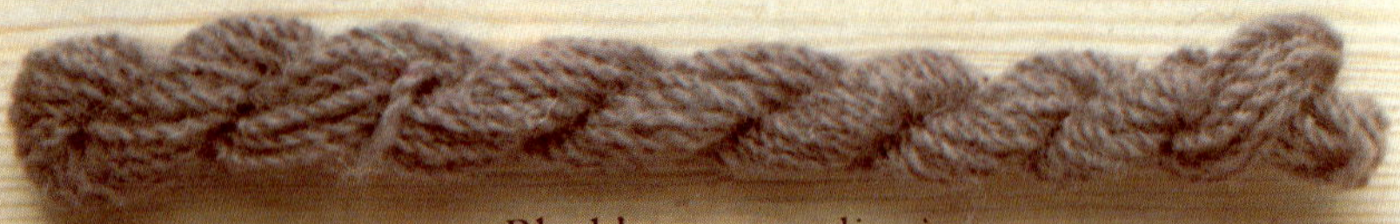

Blackberry – 1st dipping

Blackberry – 2nd dipping

Bracken using alum as a mordant

Bracken – 2nd dipping

Onion using chrome as a mordant

Onion using tin as a mordant

Walnut
Yew
Viburnum tinus
Weld
Laurel berries
Horsetail
Madder

Dyestuff	*Part of plant used*	*Source*	*Mordant*
GOLDS			
Knapweed	flowers	f	C
Marigold	flowers	f or c	A or T
Saffron	stigmas	f or c	A
Stinging nettle	flowers	f	A
Turmeric	roots	c	C
BROWNS			
Coreopsis	flowers	f	C
Kutch	wood, nuts, seed pods	c	A or none
Lichen	whole	f	None
Madder	roots (powdered)	c	C
Onion	whole	f	C
Poplar	bark	f	C
Sumach	leaves, shoots	f	None
Walnut	hulls	f or c	None or I
Yew	bark or wood shavings	f	A

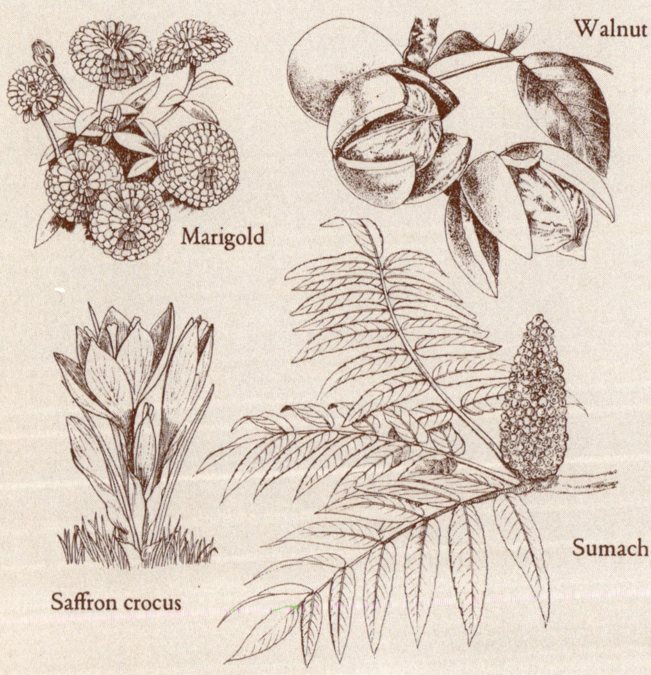

KEY

- f collect and use while fresh
- c available commercially from chemists or craft shops
- d may be collected fresh and used dried
- A mordant with alum
- C mordant with chrome
- T mordant with tin
- I mordant with iron

Dyestuff	*Part of plant used*	*Source*	*Mordant*
GREENS			
Apple	bark	f	I
Bracken	fronds, young shoots	f *	A
Coreopsis	flowers	f	I
Golden Rod	flowers	f	A
Horsetail	whole	f	A
Onion	whole	f	I
Privet	leaves	f	I
Ragwort	flowering tops	f	I
Reed	flowering tops	f	A or I
St John's Wort	flowering tops	f	A
Tomato	leaves, vines	f	A with I
Weld	whole plant	f	I

* collect in spring

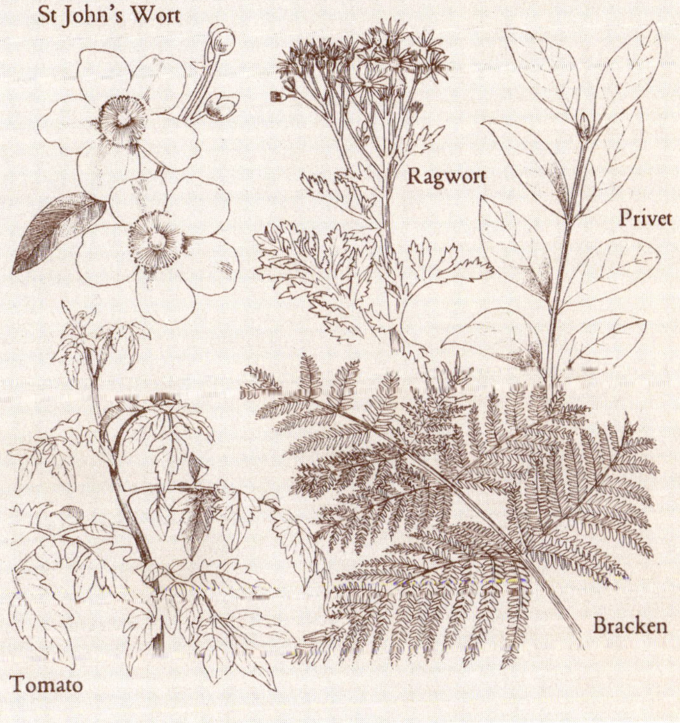

Overleaf: The pair of gloves knitted with handspun wool were dyed using the following ingredients: pine cones, lichen, onion skins, *Viburnum tinus* and young shoots of blackberry

Dyestuff	*Part of plant used*	*Source*	*Mordant*
REDS			
Bedstraw	roots	f	A or C
Cochineal	whole insects	c	A or T
Henna	leaves	c	T
Hop	stems	f	A
Madder	roots	c	A or C
BLUES			
Alkanet	roots	f or c	None
Indigo	leaves	c	see recipe after
Logwood	bark	c	None
PURPLES			
Birch	bark	f	I
Cochineal	whole insects	c	C
Elderberry	berries	f	A *
Logwood	bark	c	A
Whortleberry	berries	f	None

* with the addition of acetic acid

GREYS			
Blackberry	young green shoots	f	I
Portugese laurel	berries	f	A
Sloe	berries	f	A

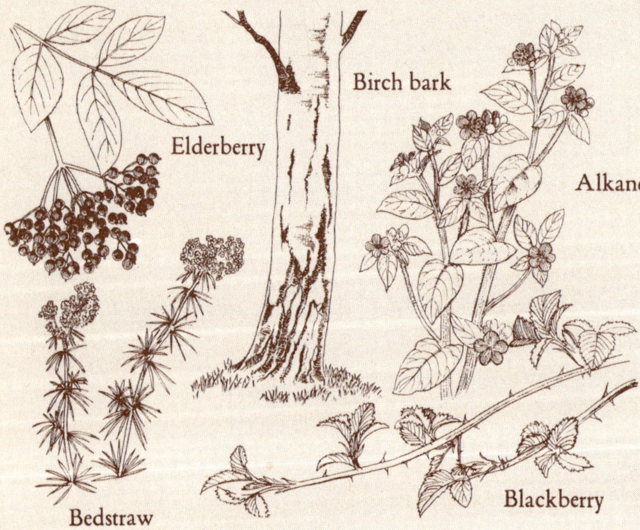

KEY

- f collect and use while fresh
- c available commercially from chemists or craft shops
- d may be collected fresh and used dried
- A mordant with alum
- C mordant with chrome
- T mordant with tin
- I mordant with iron

Method for dyeing with lichens:

You need about 0·5 kg (1 lb) of lichen to dye 0·5 kg (1 lb) of wool.

Put a layer of lichen in the bottom of the dyebath, then a layer of wool, then a layer of lichen and so on until the dyebath is almost full.

Then fill the dyebath with cold water and simmer this for some hours until the required depth of colour is reached.

What colour you get from your lichen will depend on what kind of lichen you use. Most of them produce browns, yellows, yellow-browns and oranges.

However, you may get red if you use a lichen called *Ochrolechia tartarea*, commonly found on limestone rocks. To use this, you should first bruise it and mix it with water to form a paste, then gradually add to it a mixture of 1 part of ammonia to 1 part of water.

This compound should then be left in a closed vessel in a warm place for a month or so to ferment, being stirred at regular intervals during this period.

Method for dyeing with Indigo:

Put 25 g (1 oz) of powdered indigo into a glass basin. Mix this with 225 g (8 oz) sulphuric acid, adding a little acid at a time. During this process, add about 10 g (½ oz) of powdered chalk.

You should then store this mixture in a stoppered glass jar and leave for a week before using it, during which time you should stir it at regular intervals and add a little more chalk.

When this mixture is ready, stir 10 g (½ oz) of it into the water of the dyebath, then warm it on the stove; immerse your wool and bring it slowly to the boil. Thereafter, let it simmer for about ¾ hour.

Note that such a quantity of the indigo mixture is sufficient to dye about 0·5 kg (1 lb) of wool which should not be mordanted. After dyeing, wash your wool in several lots of clean, soapless water.

Overleaf left: 'Snake Lightning' a natural dyed tapestry by Clyde Leo Olliver, using cochineal, weld, indigo and natural black

Overleaf right: A rug by Lore Youngmark using hand-spun wool in its natural colours

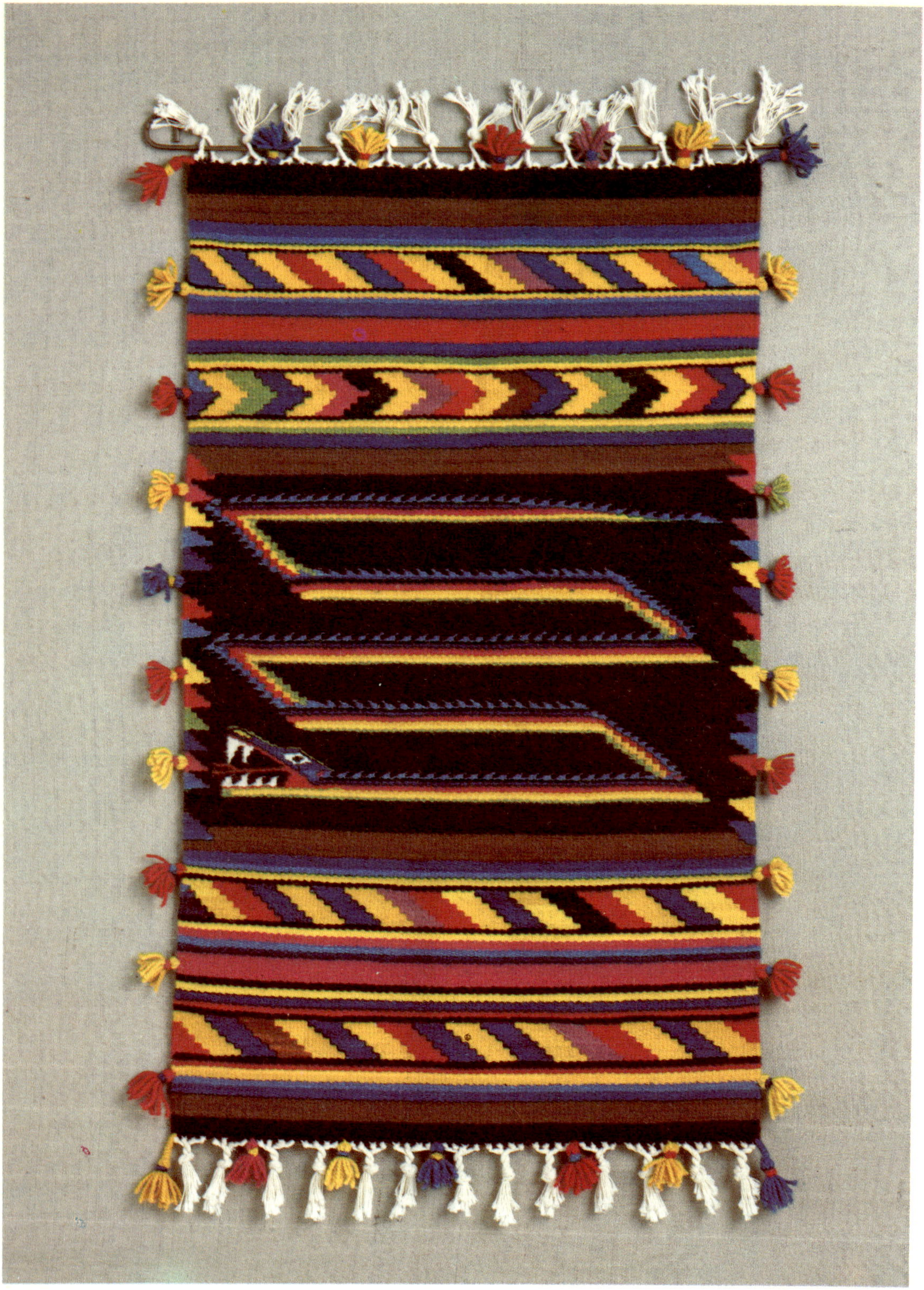

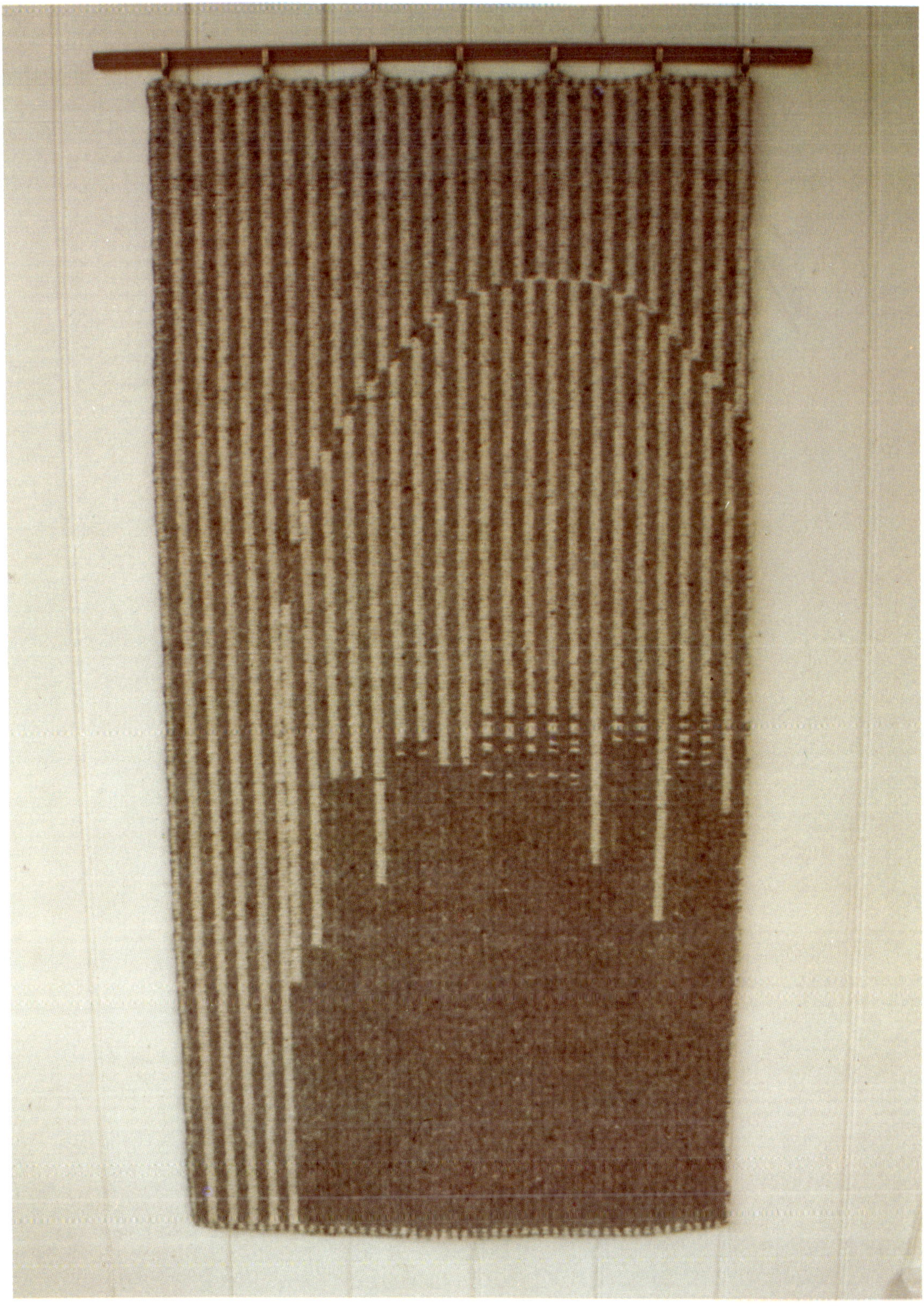

Types of Wool

The Properties of Wool

Although wool is a natural fibre, there is nothing natural about the way in which man first domesticated sheep and then bred them to become the animals we see today in our fields and on our hillsides. For all departments of animal husbandry nowadays are highly sophisticated branches of technology, and the technology that has produced the wide variety of wools and qualities of wool now available is as sophisticated as any. It is therefore to these techniques of breeding and nutrition that we owe the fact that wool now is a material with properties which make it superior in regard to the manufacture of textiles to any synthetic fibre.

So what are these properties? What in the first place is different about wool and, in the second place, how does this make it superior? Firstly it is able to absorb up to 35% of its own weight of water without ever feeling wet, whereas a comparable figure for any synthetic is only 5%. Besides, wool, when it is absorbing moisture, gives out heat. 1·5 kg (3 lb) of wool absorbs 225 g (8 oz) of water vapour over a period of several hours, giving out sufficient heat to boil away nearly 275 ml ($\frac{1}{2}$ pint) of water. Synthetic fibres, on the other hand, cannot resist chilling in this way because they do not give out any heat when they are absorbing moisture. Thus wool does not feel cold even when it is wet and, unlike synthetic fibres which allow the quantity of water on their wearer's skin to increase, by virtue of its absorbency wool keeps this constant.

In addition, wool is a very good insulator. On the sheep's back, it traps sufficient air between its fibres to keep the animal both warm in winter and cool in summer. Because of the unique structure of the individual woollen fibres, woollen yarn can contain a considerable amount of trapped air. Indeed, even in the most tightly spun worsted yarn, air comprises 60% of its volume.

Finally wool is both resilient and durable—although the breaking strength of steel, for example, is twice that of a woollen fibre, wool is about five times as elastic and therefore can be stretched and bent to the extent that one can say in fact that it is twice as tough as steel.

Where to Obtain Wool for Handspinning

Where you obtain your wool depends partly on where you live. Different countries organise their woollen trade in different ways although, equally clearly, there are similarities.

In Britain, for example, almost all wool produced in the country is destined eventually for the central collecting house of the British Wool Marketing Board in Bradford. However, there are subsidiary collecting stations where you may obtain wool, dotted about the country.

However, there are exceptions to this rule and one of these is where a farmer keeps no more than four animals in a single flock. Then he can sell the fleeces from these directly to the public. In addition, if he keeps a specialised breed of sheep, he can obtain a special exemption from this rule from the Marketing Board and again may sell his wool directly to you.

Alternatively, you can keep your own sheep for their wool as some spinners do, though this isn't to say that the quality of the yarn you spin will reflect your trouble. For clearly, if you do, you will never be able to offer yourself the wide variety of choice of fleeces obtainable from the collecting stations of the British Wool Marketing Board, for example. Besides, as sheep grow older, so the quality of their fleeces deteriorates. Consequently, if you choose to produce your own wool, you will find yourself having to dispose of the fleeces of your older animals in some other way than by spinning them, whereas at the Wool Marketing Board, each year you may have a choice of the fleeces of a number of healthy shearlings.

As to colour too, at the Wool Marketing Board, you will quite naturally have the chance to select a fleece from a much wider range of coloured fleeces, whereas with your own sheep, you will be tied to whatever colours of fleece nature in her whimsical way produces.

The Availability of Fleeces

However, though the means of collection and distribution of wool may vary little from country to country, this is not to say that the types of wool are equally constant. For example, in Australia, you might find it considerably more difficult to acquire a fleece from one of those breeds of sheep which were the first to be taken to that continent from

Previous page: Herdwick sheep found in the fells of north-west England

Britain and which are still common in Britain; though you would have little trouble in obtaining that of a Merino. Similarly, in Britain you might expect to have difficulty in obtaining a Merino fleece because very few Merinos are kept in Britain whereas you might well be embarrassed at the plethora of fleeces offered to you coming from traditionally English breeds of the animal.

However, at least in Britain, New Zealand, Australia and North America, somehow you should be able to acquire the fleeces of more or less the same breeds of sheep, if only because of the historical connections between the agricultural traditions of these countries. Indeed, in theory at least, the only difference between each of these countries should lie in the degree of difficulty you have in obtaining a particular type of fleece. Thus, while Merinos predominate in Australia, the traditionally English breeds of sheep which preceded them are still kept there.

Meanwhile, in New Zealand, you will find it easier to obtain a Merino fleece than you would in Britain because there was a time when Merinos were as popular in New Zealand as in Australia. However, now it seems that the Romney predominates there because of the profitability of New Zealand lamb, though by no means to the total exclusion of other well-known breeds.

Again in the USA, most of those breeds which originally went there from Britain still exist there, though new breeds such as the Romeldale have been bred from these and other breeds, just as the Polwarth and the Corriedale were newly bred in New Zealand.

A jacob ram displaying his piebald fleece

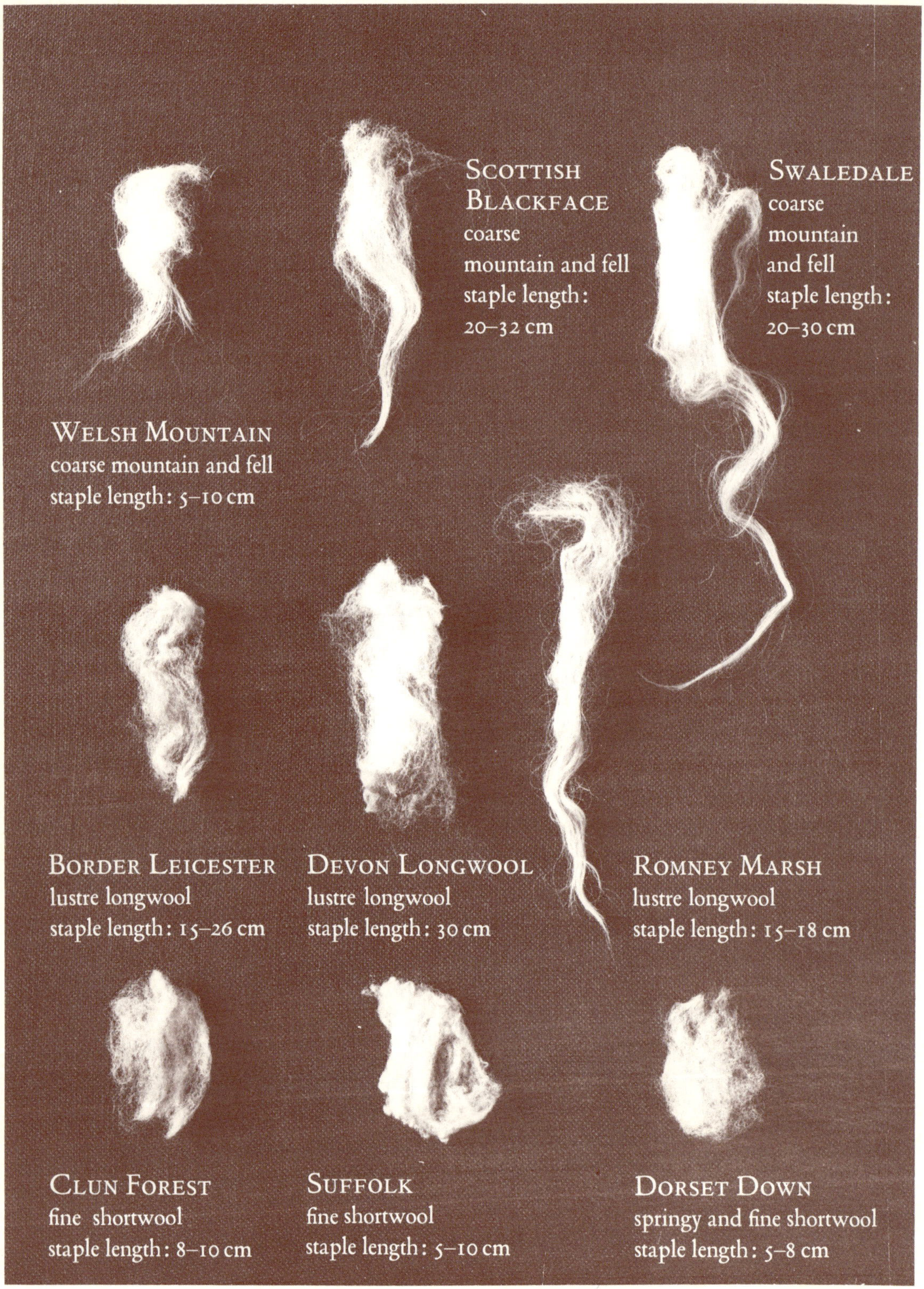
WELSH MOUNTAIN
coarse mountain and fell
staple length: 5–10 cm
SCOTTISH BLACKFACE
coarse
mountain and fell
staple length:
20–32 cm
SWALEDALE
coarse
mountain
and fell
staple length:
20–30 cm
BORDER LEICESTER
lustre longwool
staple length: 15–26 cm
DEVON LONGWOOL
lustre longwool
staple length: 30 cm
ROMNEY MARSH
lustre longwool
staple length: 15–18 cm
CLUN FOREST
fine shortwool
staple length: 8–10 cm
SUFFOLK
fine shortwool
staple length: 5–10 cm
DORSET DOWN
springy and fine shortwool
staple length: 5–8 cm

Different Types of Wool

Wherever you live and whatever types of fleece you may be able to get, broadly you will find that these fall into three categories: longwool, medium wool and shortwool. It is broadly true to say that the long-woolled varieties of sheep produce a coarser wool than that of the short-woolled types, while those with wool of a medium length produce wool which is also of a moderate coarseness.

However, this needs some qualification. The Devon Longwool, for example, whose fibres have a staple length of 30 cm (12 in), also produces wool that is very coarse; yet this is not as coarse as that of the Herdwick whose fibres have a staple length of only 13 cm (5 in). Similarly, the breeds with the longest wool, the Lincoln and the Wensleydale, with a staple length of between 30 and 40 cm (12 and 16 in), are certainly not as coarse in the fibre as the Herdwick. Again, the Merino which produces the finest wool in the world has an average staple length of between 5 and 10 cm (2 and 4 in), whereas the Southdown, for example, produces wool whose fibres, while they measure only 2 to 6 cm (1 to $2\frac{1}{2}$ in) in length, are somewhat coarser.

Wool Suitable for Handspinning

However, within the three main categories of fleeces, the fleeces of different breeds of sheep have different characteristics. Therefore it is worth knowing which are curly and which are straight in the fibre; which are lustrous and which are dull and which are likely to be easy to prepare and spin and which are not.

Beyond this though, as far as one can pass any opinion on what is the most suitable wool for the handspinner, it is only to say that one with a staple length of between 10 and 20 cm (4 and 8 in) is easier to spin than one whose fibres are either shorter or longer. A wool whose fibres are crimpy is easier to spin than one whose fibres are straight. Apart from this, the only other aspect to consider is the purpose for which you intend to use your yarn when you have spun it.

Thus, it is not advisable to spin yarn to knit into a vest from wool which is really too coarse to make into anything except carpets. By the same token, for most knitwear you would be wise to choose a wool whose fibres are fairly fine and soft and again, if you want to dye it, white in colour. However, the entire fleece need not be white. If you want to include either white or dyed wool in a garment, contrasted with a darker wool, you could get both from a piebald fleece.

Opposite: Different breeds of sheep have wool of different quality and staple length

No doubt the desire to do exactly this and the fact that, in other ways too, it can match up to the handspinner's requirements, account for the increasing popularity of the Jacob. However, here I should sound a note of caution. The quality of the fleeces of the Jacob is more variable than those of most other breeds. Perhaps this is because, kept more for aesthetic than commercial reasons, flocks of Jacobs have been far-flung and disparate. Therefore, breeding has generally taken place within a single flock, tending to exaggerate characteristics in the wool of these individual flocks in one or another direction. Consequently, unless you particularly seek a fleece with a harsher handle, you should take care to avoid the coarser examples of the fleece of the Jacob.

By comparison, the fleece of the Suffolk, for instance, is more reliable though a little more difficult to spin, its fibres having a staple length of only 5 to 10 cm (2 to 4 in), compared with 8 to 13 cm (3 to 6 in) in the Jacob. Again, the Shetland produces a fleece of similar softness to the Suffolk and, with a staple length of 10 to 13 cm (4 to 5 in) is easier to spin. Moreover, it comes in a variety of colours ranging from a reddish-brown known as moorit (moor-red) through browns and greys to fawn and white. However, this is not so easy to obtain, even in Britain where it has its origins, as few flocks of it are ever maintained.

At any rate, what follows is a list of breeds of sheep whose wool I consider to be both easy to spin and suitable when spun for knitting and crochet. On the other hand, if you want yarn with which to weave, you may seek a fleece whose fibres are both longer and coarser, particularly for use for the warp threads. If you want to weave tweeds, for example, you may seek one with what in all other cases may be an undesirable characteristic: the presence of kemps (hairy, undyeable fibres). It is these in part which account for the distinctive appearance of tweeds from other textiles. Such a fleece is that of the North Country Cheviot.

Kerry Hill

Fleeces most suitable for handspinning into knitting yarn

	*Quality count	Staple length cm	Staple length in
Cheviot	50–56's	10	4
Clun Forest	56–58's	8–10	3–4
Corriedale	50–58's	15–20	6–8
Dorset Horn	54–58's	8–10	3–4
Jacob	44–56's	8–15	3–6
Kerry Hill	56–58's	10	4
Polwarth	58–64's	9–11	$3\frac{1}{2}$–$4\frac{1}{2}$
Portland	54–58's	8–10	3–4
Shetland	56–60's	10	4
Shropshire	50–58's	9–11	$3\frac{1}{2}$–$4\frac{1}{2}$
Suffolk	54–58's	5–10	2–4

Suffolk

Cheviot

Clun Forest

Dorset Horn

* *Quality count:* Though different means of classifying the quality of wools are employed in different countries, in the United Kingdom at least, this is done according to the Bradford Quality Count.

According to this, the figures given in the columns above refer to the number of hanks, 304 m (560 yds) long, which can be spun from one pound of a particular kind of wool. Naturally this follows that a finer wool produces a greater length of wool per pound than a coarser one and therefore that a finer wool has a higher quality number.

By way of comparison, the other most common breeds of sheep with their characteristics are as follows:

Longwools

	Quality count	Staple length	
		cm	*in*
Border Leicester	40–46's	15–26	6–10
Cotswold	44–46's	18–26	7–10
Devon Longwool	32–36's	30	12
Romney Marsh	48–56's	15–18	6–7
Leicester	40–46's	30–35	12–14
Lincoln Longwool	38–44's	30–40	12–16
South Devon	36–40's	18	7
Teeswater	40–48's	30	12
Wensleydale	40–48's	35	14
Whiteface Dartmoor	30–40's	26–30	10–12

Romney

Border Leicester

SHORTWOOLS OR DOWN BREEDS			
Devon Closewool	46–50's	8–10	3–4
Dorset Down	56–58's	5–8	2–3
Hampshire Down	56–58's	5–9	2–3½
Merino	65–100's	5–10	2–4
Orkney	50–56's	4–8	1½–3
Oxford Down	50–58's	15	6
Radnor	50–58's	8–18	3–7
Ryeland	56–58's	8–10	3–4
Southdown	56–60's	2–6	1–2½

Dorset Down

MOUNTAIN OR FELL (COARSE-WOOLLED) BREEDS			
Dalesbred	32–40's	20	8
Derbyshire Gritstone	50–56's	15–20	6–8
Exmoor Horn	50–56's	8–10	3–4
Herdwick	28–32's	13	5
Lonk	32–44's	10–15	4–6
North Country Cheviot	50–56's	5–8	2–3
Rough Fell	32–36's	20	8
Scottish Blackface	28–32's	20–32	8–13
Swaledale	28–32's	20–30	8–12
Welsh Mountain	36–50's	5–10	2–4
Black Welsh Mountain	36–50's	10	4

Black Welsh Mountain

Choosing a Fleece

Having decided in general what type of fleece you want, perhaps even the exact breed of sheep from which it will come, you will still need to know what points to examine when a selection of fleeces from such animals is presented to you. What this amounts to is testing their condition.

For when such a variety is available to you, in the first place you should be able to obtain one that is clean. This does not mean that you should be too fussy – pieces of dung stuck to the hind-quarters of the fleece are an occupational hazard of the handspinner – only that there is no reason why you should accept fleeces from animals that have obviously rolled in a peat bog and suffered accordingly. This only makes more work for you later.

Similarly, when you examine a fleece, you should test its fibres. The health of the animal from which it has been shorn is reflected in these fibres. Thus an illness or period of malnutrition during the course of the year in which it has been growing, makes them brittle at that part of the fibres which was growing when the animal was sick or starved. This brittleness is caused by the attenuation of the fibres at such a time so that this part of them is in fact somewhat thinner than the rest of them.

To test the fibres in a fleece then, take a few strands of fibre, loop them around the index finger of one hand, gripping the ends of the strands between thumb and third finger. Then with the strands tightly stretched, flick them with the fingers of your other hand. If they break or begin to disintegrate this may be a sign that they have come from a sick or badly nourished animal. In any case, it is an indication that they are unsuitable for handspinning.

However, you need not necessarily rely on your own judgement in this. If, for example, you buy your wool from an establishment such as the British Wool Marketing Board, you are likely to find that its representative is an expert judge in such matters and will be willing to give you the benefit of his or her experience.

Finally, in choosing your fleece, there are three kinds of fleece to avoid. The first of these is a badly matted fleece. This will be extremely difficult to tease out, impossible to card and will probably be unusable. The second is the fleece containing substantial quantities of either kemps or hair. You can recognise kemps by the fact that they are short, thick, wavy fibres with pointed tips and roots. These are shed periodically into the fleece and also have a white, chalky appearance. You should avoid them because they do not absorb dyes, although they can be useful for the purpose of achieving special effects for this very reason.

Fleeces containing hair on the other hand, produce a very coarse yarn. As the fleeces of primitive sheep were a mixture of both wool and hair, some fleeces still contain this. Therefore, unless you want to coarsen your yarn with it, you should avoid these also.

The third type of fleece you should not choose is one which has been shorn too late. A fleece stops growing in the winter, then starts again in the spring. As it is easier to clip it shortly after it has started to grow again, this is the usual practice. The very small amount it is allowed to grow before clipping is called the 'rise'. If clipping is delayed too long, this new growth achieves a measureable length which presents itself on the underside of the fleece in the form of short, fluffy ends to the fibres. These will make your wool more difficult to spin and produce a yarn of a similarly fluffy, but also uneven, tufty texture.

Of course, if this is what you want then there is no reason to avoid such a fleece. However, ideally, particularly if you are just learning to spin, what you want is the fleece of a one-year-old sheep, called a hog or hoggett, which has been shorn at the right time by an expert.

The Patterns

INTRODUCTION

The following section comprises a selection of patterns for garments that are easy to make and which also, I hope, will be of particular appeal to the handspinner.

The emphasis is on simplicity. In addition I have included one or two items which do not require much wool and which should not be too daunting to the beginner.

Indeed, at least one of these, the cushion, owes its charm in part to the use of small quantities of wool in a variety of textures. It is therefore probably a good pattern to work on at first as it does not require a large quantity of uniformly spun yarn in its manufacture.

However, the patterns that follow merely represent a basic range of items which are particularly useful because, in most cases, they are wearable – the kind of things, indeed, for whose manufacture Man invented handspinning in the first place. They by no means represent the complete range of types of articles you can make in this way nor the pinnacle of what you can achieve from a creative point of view, either by weaving, knitting or crochet.

Indeed, in any of these ways, you can add to or vary what I have done, both by reference to the individuality of your own handspun yarn and by turning to different ways of using it.

At one extreme, for example, you could use it to weave wall handings. This, after all, probably offers the widest scope for the exercise of the imagination. In addition, you can use simple weaving techniques to create small items, such as table mats out of small quantities of handspun yarn.

TENSION

As you create your own handspun yarn, you will soon realise by referring to what other handspinners produce, that all your yarns are different. Unlike machine-spun yarns, none of them are to a standard or are likely to be uniform in thickness.

This means that no two handspinners, knitting or crocheting their own yarn, will create the same tension between stitches.

First then, before embarking on an entire item, you should knit a tension square to determine what is the correct tension to use for your own yarn. You may then have to vary the size of the needles you use or adapt the size of garment given.

Consequently you should only take my instructions in this respect as a guide, as indeed you should when working to any commercial pattern.

Illustrated here are two tension squares knitted on the same size needles, using commercially available two-ply knitting yarn in one (right) and my equivalent handspun yarn in the other (left).

Quantities of wool required

When attempting to make a larger item, it is best first to spin all the wool you require.

This eliminates the frustration of having to start spinning again halfway through making a garment in order to finish it. In additon, it also makes it more likely that the yarn you spin will be of a uniform gauge as this does tend to vary from 'sitting' to 'sitting'.

However the exact weight of wool you will require for any particular garment raises another problem as you will find that no two handspun yarns are of the same density. Therefore once again you will discover that the figures I have given in the patterns that follow can be no more than a guide.

The same applies to commercial patterns. In this regard, it is perhaps worth mentioning that handspun yarn is generally lighter in weight than its commercially produced equivalent.

One point that is important is to knit garments out of plyed yarn. Those knitted using single ply yarn may tend to curl and twist, thus spoiling the shape of the garment.

Herdwick Sweater

Measurements
given in inches
(centimetres in brackets)
To fit chest size

38 (97)	40 (102)	42 (107)
Length		
26 (66)	27 (66)	$26\frac{1}{2}$ (68)
Sleeve seam		
18 (46)	18 (46)	18 (46)

Weight of wool: 24 oz (675 g)

Tension: Work at a tension of 11 stitches and 15 rows to $2\frac{1}{2} \times 2\frac{1}{2}$ inches (5 × 5 cms) measured over stocking stitch using No. 8 (4 mm) needles.

Abbreviations: K – knit plain; p – purl; st – stitch; dec. – decrease; inc. – increase; dec. – decrease; ins – inches; cms – centimetres.

Use one pair No. 8 (4 mm) knitting needles and one pair No. 10 ($3\frac{1}{4}$ mm).

THE BACK

Using No. 10 ($3\frac{1}{4}$ mm) needles cast on 102 (106) (110) sts and work in double rib as follows:
1st row: K 2, p 2, repeat to end of row.
2nd row: P 2, k 2, repeat to end of row.

Repeat last two rows until work measures $2\frac{3}{4}$ ins (7 cms).

Change to No. 8 (4 mm) needles and continue in the double rib until back measures $18\frac{1}{2}$ ins (47 cms) from the beginning. End with right side facing.
For armholes: Cast off 3 sts at beginning of next 2 rows, then dec. 1 st at each end of next and every alternate row until 78 (82) (86) sts remain.**

Work until back measures 26 (26) ($26\frac{1}{2}$) ins, 66 (66) (68) cms, from beginning ending with right side facing.
For shoulders: Cast off 7 (8) (9) sts at beg. of next two rows, then 7 sts at beg. of next 4 rows.

Break wool and leave remaining 36 (38) (40) sts on a spare needle.

THE FRONT

Work as for back to **. Then continue in the rib pattern and work straight until front measures 24 (24) ($24\frac{1}{2}$) ins, 61 (61) (63) cms, ending with right side facing.
To divide for neck: Rib 30 (32) (34) sts, turn and leave remaining sts on spare needle.
The left front shoulder: Work on these 30 (32) (34) sts. Dec. 1 st at neck edge on every row until 21 (23) (25) sts remain.

Work straight until front matches back to beginning of shoulder shaping, ending at the side edge.
To slope the shoulder: Cast off 7 (8) (9) sts at beginning of next row. Work 1 row; then cast off 7 (8) (9) sts at the beginning of the next row. Work 1 row and then cast off remaining 7 sts.
The right front shoulder: With right side facing, slip centre 18 sts onto spare needle, rejoin wool to inner end of remaining sts and work as given for left front shoulder.

THE SLEEVES

(Two alike)
Using No. 10 ($3\frac{1}{4}$ mm) needles, cast on 46 (50) (54) sts. Work $2\frac{3}{4}$ ins (7 cms) in k 2, p 2 rib.

Change to No. 8 (4 mm) needles and continue in rib, inc. 1 st at each end of 7th and every subsequent 7th row until there are 70 (74) (78) sts, taking the extra sts into the rib.

Work until sleeve seam measures 18 ins (46 cms) ending with right side facing.
To shape top: Cast off 3 sts at beginning of next 2 rows, then dec. 1 st at each end of next and every alternate row until 38 (40) (42) sts remain.

Work 1 row, then dec. 1 st at each end of every row until 18 sts remain. Cast off loosely in rib.

NECKBAND

Join right shoulder seam. With right side facing, rejoin wool and using No. 10 ($3\frac{1}{4}$ mm) needles, pick up and k 20 sts down left side of neck, k 18 sts from centre, pick up and k 20 sts from right side of neck, then k 36 (38) (40) sts across back. [94 (96) (98) sts].

Work 2 ins (5 cms) in k 2, p 2 rib. Cast off evenly, using a larger size needle.

TO MAKE UP

Press lightly. Join left shoulder seam continuing across the neck band. Join side and sleeve seams. Insert sleeves into armholes. Fold neck band in half and catch down to wrong side.

Opposite: A heavyweight, outdoor sweater knitted out of the natural grey wool of the Herdwick sheep

Man's Sleeveless Pullover

Measurements
given in inches
(centimetres in brackets)
To fit chest size

38 (97)	40 (102)	42 (107)
Length		
24 (61)	24½ (62)	25 (64)

Weight of wool: 10 oz (285 g)

Tension: Work at a tension of 7 stitches and 12 rows to 1 × 1 inches (2.5 × 2.5 cms) measured over stocking stitch using No. 10 (3¼ mm) needles.

Abbreviations: K – knit plain; p – purl; st – stitch; inc. – increase; dec. – decrease; tog. – together; st st – stocking stitch; ins – inches; cms – centimetres.

Note: The instructions are given for the 38-in (97-cm) size. Where they vary, work the figure in the first bracket for the 40-in (102-cm) size and the figure in the second bracket for the 42-in (107-cm) size.

Use one pair of No. 12 (2½ mm) and one pair of No. 10 (3¼ mm) needles.

THE FRONT

Using No. 12 (2½ mm) needles, cast on 120 (124) (128) sts. Work in single rib (k 1, p 1) for 3 ins (7.5 cms).

Next row: Rib 12 (14) (16) sts (inc. in next st, rib 5) 16 times, rib 12 (14) (16) sts [136 (140) (144) sts].

Change to No. 10 (3¼ mm) needles and work in pattern as follows:

1st row: K 3, p 1, repeat to end.

2nd row: K 1, p 3, repeat to end.

Repeat these 2 rows until work measures 15 (15½) (16) ins, 38 (39) (41) cms, from beginning ending with a 2nd pattern row.

To shape armholes and divide for neck: Cast off 7 sts at beginning of next 2 rows.

Next row: Work 2 tog., rib 59 (61) (63) sts; turn, slip remaining sts onto a stitchholder and leave.

Keeping pattern correct, dec. 1 st at neck edge at beginning of next and every following 3rd row, at the same time dec. 1 st at armhole edge on every alternate row until 6 more decreases (7 in all) have been worked at armhole edge.

Proceed with further decreases at armhole edge but still decrease at neck edge on every 3rd row as before until 33 (35) (37) sts remain.

Continue on these sts until work measures 9 (9) (9½) ins, 23 (23) (24) cms, from beginning of armhole shaping, finishing at armhole edge.

To shape shoulder: Cast off 11 (12) (13) sts at beginning of the next and following alternate row.

Work 1 row. Cast off remaining 11 sts.

With right side of work facing, rejoin wool to remaining sts, pattern to end of row, k 2 tog.

Keeping pattern correct, dec. 1 st at neck edge of next row and every following 3rd row, at the same time dec. 1 st at the armhole edge on every alternate row.

Complete to match first side.

THE BACK

Work as front until armhole shaping is reached.

To shape armhole: Cast off 7 sts at beginning of next 2 rows. Dec. 1 st at both ends of next and every alternate row until 108 (112) (116) sts remain.

Continue on these sts until work measures same as front up to shoulder shaping.

To shape shoulder: Cast off 11 (12) (13) sts at beginning of next 4 rows. Then cast off 11 sts at beginning of next 2 rows. Cast off.

NECKBAND

Join right shoulder seam.

Using No. 12 (2½ mm) needles and with right side of work facing, begin at top of left front neck and k up 87 (87) (91) sts to centre V. K up to 1 st from centre V (by picking up loop of centre V and knitting into back of it). K up 88 (88) (92) sts from right front neck and finally k up 42 sts across centre back [218 (218) (226) sts].

Work 9 rows in k 1, p 1 rib, dec. 1 st each side of the st at the V on every row.

Cast off in rib. Join left shoulder and neckband seam.

ARMBANDS

Using No. 12 (2½ mm) needles, with right side of work facing, k up 182 (182) (190) sts round armhole.

Work 9 rows in k 1, p 1 rib. Cast off.

TO MAKE UP

Omitting ribbing, with wrong side of work facing, block fabric by pinning out round edges. Press with a warm iron and damp cloth. Using back st, join side seams.

Opposite: This slipover uses natural black wool from the Jacob sheep

Child's Cardigan

Measurements
given in inches
(centimetres in brackets)

To fit chest size		
20 (51)	22 (56)	24 (61)
Sleeve seam		
6 (15)	6½ (16)	7 (18)
Length		
11 (28)	11½ (29)	12 (31)
Side seam		
6 (15)	6½ (16)	7 (18)

Weight of wool: 2½ oz (70 g)

Tension: Work at a tension of 11 stitches and 15 rows to 2 × 2 ins (5 × 5 cms) on No. 9 (3½ mm) needles measured over stocking stitch.

Abbreviations: K – knit plain; p – purl; st – stitch; st st – stocking stitch; beg. – beginning; dec. – decrease; inc. – increase; dc – double crochet; ins – inches; cms – centimetres.

Use No. 9 (3½ mm) needles.

Note: The instructions are given for the 20-in (51-cm) size. Where they vary, work the figure in the first bracket for the 22-in (56-cm) size and the figure in the second bracket for the 24-in (61-cm) size.

THE BACK

Using No. 11 (3 mm) needles, cast on 60 (66) (72) sts. Work in k 1, p 1 rib for 2 ins (5 cms).

Change to No. 9 (3½ mm) needles and beg. with a k row work in st st until back measures 6 ins (15 cms) from beginning, ending with right side facing.

To shape the armholes: Cast off 2 (3) (3) sts at beginning of next 2 rows.

Dec. 1 st at each end of next and every alternate row until 20 sts remain. Cast off.

THE RIGHT FRONT

Using No. 11 (3 mm) needles, cast on 30 (34) (38) sts. Work in k 1, p 1 rib for 2 ins (5 cms).

Change to No. 9 (3½ mm) needles and beg. with a k row continue in st st until work measures 6 (6½) (7) ins, 15 (16) (18) cms from beginning ending with wrong side facing.**

To shape the armhole: Cast off 2 (3) (3) sts, p to end.

*** Dec. 1 st at end of next row and every alternate row until 15 (16) (17) sts remain, ending at front edge.

To shape the neck: Cast off 5 (6) (7) sts, k to last 3 sts, k 2 tog., k 1.

Dec. 1 st at armhole edge on every alternate row as before. At the same time, dec. 1 st at neck edge on every row until 3 dec. have been worked at neck edge. Continue decreasing at armhole edge until 1 st remains. Fasten off.

THE LEFT FRONT

Work as given for right front to ** but ending with the right side facing.

To shape the armhole: Cast off 2 (3) (3) sts, k to end.

Complete as for right front from *** to end.

SLEEVES

(Two alike)

Using No. 11 (3 mm) needles cast on 34 (38) (42) sts. Work 2 ins (5 cms) in k 1, p 1 rib.

Change to No. 9 (3½ mm) needles and beg. with a k row work in st st, inc. 1 st at each end of 5th and every subsequent 4th row until there are 50 (54) (58) sts.

Continue in st st until sleeve measures 6 (6½) (7) ins, 15 (16) (18) cms, ending with right side facing.

To shape sleeve top: Cast off 2 (3) (3) sts at beginning of next 2 rows. Dec. 1 st at each end of next and every alternate row until 8 sts remain. Cast off.

TO MAKE UP

Press lightly. Set in sleeves and then join side and sleeve seams. Crochet one row of dc along both front and neck edge. Crochet one row of shells along both fronts and neck edge. Sew on buttons.

Above and opposite: Both the cardigan and the trim use undyed wool from Shetland sheep

Bootees

Weight of wool: ½ oz (14 g)

Tension: Work at a tension of 7 stitches and 12 rows to 1 inch (2.5 cms) measured over garter stitch using No. 10 (3¼ mm) needles.

Abbreviations: K – knit plain; p – purl; st – stitch; g st – garter st; w fd – wool forward; tog. – together; sl — slip; st st – stocking stitch; dec. – decrease; ins – inches; cms – centimetres.

Use No. 10 (3¼ mm) knitting needles.

METHOD

(both alike)

Using No. 10 (3¼ mm) needles cast on 41 sts. Work 4 rows g st (k every row). Change to pattern as follows:

1st row: Sl 1, k to end of row.

2nd row: Sl 1, *p 1, k 1; repeat from * to end of row.

Repeat these two rows until work measures 2½ ins (6 cms) ending with a 2nd pattern row.

Slot row: K 2, *w fd, k 2 tog.; repeat from * to last st, k 1.

Next row: All p.

Next row: K 27, turn, slipping remaining 14 sts onto spare needle unworked.

K 13, turn, slipping remaining 14 sts onto spare needle unworked.

Work on centre 13 sts for 14 rows in g st. Break off wool.

Next row: With right side of work facing, rejoin wool at side edge and k 14 sts from spare needle, pick up and k 12 sts from row ends of one side of instep, k 13 sts from centre, pick up and k 12 sts from row ends of other side of instep and then k across remaining 14 sts (65 sts).

Work 7 rows in g st.

TO SHAPE THE HEEL AND TOE

1st row: K 1, k 2 tog., k 22, k 2 tog., k 11, k 2 tog., k 22, k 2 tog., k 1.

2nd row: All k.

3rd row: K 1, k 2 tog., k 21, k 2 tog., k 9, k 2 tog., k 21, k 2 tog., k 1.

4th row: All k.

5th row: K 1, k 2 tog., k 20, k 2 tog., k 7, k 2 tog., k 20, k 2 tog., k 1.

6th row: All k.

Cast off remaining 53 sts loosely. Make a matching bootee in the same way.

TO MAKE UP

Press lightly. Sew up underfoot and leg seams. Make twisted cords 16 ins (40 cms) long and thread through the slot rows of each bootee.

Mittens

Weight of wool: ½ oz (14 g)

For tension and abbreviations, see under Bootees.

METHOD

(Both alike)

Using No. 10 (3¼ mm) needles, cast on 35 sts. Work 4 rows in g st. Continue in pattern as follows:

1st row: Sl 1, k to end of row.

2nd row: Sl 1 * p 1, k 1; repeat from * to end of row.

Repeat these two rows until work measures 2 ins (5 cms) ending with a 2nd pattern row.

Slot row: K 1, *w fd, k 2 tog.; repeat from * to end.

Next row: All p.

Work 20 rows in st st.

TO SHAPE THE TOP

1st row: K 1, *k 2 tog., k 3; repeat from * to last 4 sts; k 2 tog., k 2.

2nd row: All p.

3rd row: K 1, *k 2 tog., k 2; repeat from * to last 3 sts; k 2 tog., k 1.

4th row: All p.

5th row: K 1, *k 2 tog., k 1; repeat from * to last 2 sts; k 2 tog.

6th row: All p.

7th row: K 2 tog. 7 times. Break off wool leaving a long end and thread this through the remaining 7 sts. Draw up tightly and fasten off securely. Make a matching mitten in the same way.

TO MAKE UP

Press lightly. Join side seams. Make twisted cords 16 ins (40 cms) long and thread through slot rows of both mittens.

Opposite: Undyed yarn spun from the Jacob was used to knit these baby garments

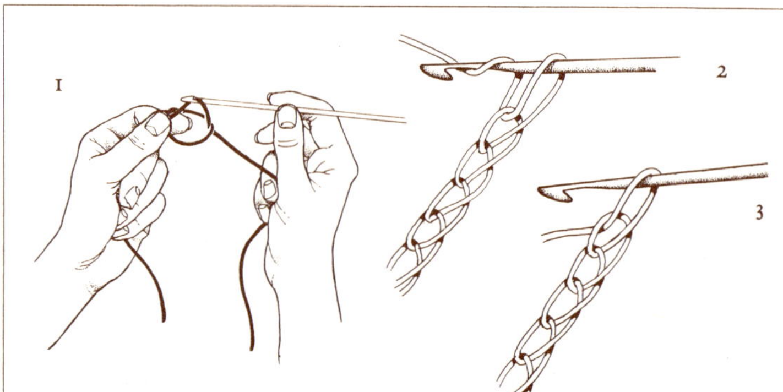

Cords

1. To start the cord, make a loop in the wool and pull the thread through using a crochet hook
2. Catch the thread with the hook
3. Pull it through the loop so forming a chain

Bonnet

Weight of wool: ½ oz (14 g)

Tension and abbreviations as for Bootees.

METHOD

Using No. 10 (3¼ mm) needles begin at the face edge by casting on 72 sts. Work 4 rows in g st. Continue in pattern as follows:

1st row: Sl 1, k to end of row.

2nd row: Sl 1, *p 1, k 1; repeat from * to end of row.

Repeat these 2 rows until work measures 3 ins (7.5 cms) from beginning.

Starting with a k row, work 4 rows in st st.

TO SHAPE THE CROWN

1st (decrease) row: (K 2 tog., k 10) 6 times.

2nd row: All p.

Repeat last 2 rows 9 times more, working 1 st less between dec. on each successive repeat of the decrease row until 12 sts remain.

Next row: K 2 tog. 6 times. Break off wool leaving a long end and thread this through the remaining 6 sts. Draw up rightly and fasten off securely.

TO MAKE UP

Press lightly. Join seam from crown to beginning of pattern. Sew an 18-in (46-cm) length of ribbon to each side of the bonnet, making small loops at the bonnet edge.

Ladies' Waistcoat

Measurements
given in inches
(centimetres in brackets)
To fit bust size
34–36 (86–91)
Length to shoulder
19 (48)

Weight of wool: 5½ oz (155 g)

Tension: Work at a tension of 12 stitches and 17 rows to 2 × 2 inches (5 × 5 cms) measured over stocking stitch using No. 8 (4 mm) needles.

Abbreviations: K – knit plain; p – purl; st – stitch; beg. – beginning; dec. – decrease; s s – slip stitch; st st – stocking stitch; dc – double crochet; ins – inches; cms – centimetres.

Use No. 8 (4 mm) needles.

THE BACK

Using No. 8 (4 mm) needles cast on 94 sts. Work in st st for 7 rows.

Next row (hemline): All k.

Starting with a k row, continue in st st for 12 ins (30 cms) measured from the hemline row.

To shape armholes: Cast off 4 sts at beg. of next 2 rows. Dec. 1 st at each end of every alternate row until 66 sts remain.

Continue until armhole measures 7 ins (17 cms).

To shape shoulder: Cast off 6 sts at beginning of next 6 rows.

Cast off centre 30 sts.

THE LEFT FRONT

Using No. 8 (4 mm) needles cast on 44 sts. Work 7 rows in st st.

Next row (hemline): All k.

Beginning with a k row, continue in st st until front measures 12 ins (30 cms) from hemline row ending with a p row.**

To shape armhole: Cast off 4 sts, k to end of row.

Next row: All p.

***Dec. 1 st at armhole edge on next and following alternate rows until 36 sts remain.

Continue until front edge measures 16 ins (40 cms) ending at the front edge.

To shape neck: Dec. 1 st at beginning of next and every alternate row, keeping armhole edge straight, until 18 sts remain.

Continue until front measures the same as the back at the armhole edge, ending at the armhole edge.

To slope shoulder: Cast off 6 sts at beginning of next and following 2 alternate rows.

THE RIGHT FRONT

Work as for left front until ** but ending with a k instead of a p row.

To shape armhole: Cast off 4 sts, p to end of row.

Next row: All k.

Work to match the left front from *** to end.

TO MAKE UP

Press lightly. Join side and shoulder seams. Work 2 rows of dc around armhole edge, neck edge and down both fronts. Turn up hem allowance and s s into position.

Attach 2 pairs of twisted wool ties to front of garment and apply wool embroidery as desired.

Opposite: The warm cream wool used for this waistcoat comes from Suffolk sheep

Hooded Sweater

Measurements
given in inches
(centimetres in brackets)

To fit bust size		
34 (86)	36 (91)	38 (97)
Length		
25 (64)	25 (64)	$25\frac{1}{2}$ (65)
Sleeve seam		
19 (48)	19 (48)	19 (48)

Weight of wool: 19 oz (535 g)

Tension: Work at a tension of 9 stitches and 12 rows to measure 2 × 2 in (5 × 5 cm) measured over stocking stitch using No. 7 ($4\frac{1}{2}$ mm) needles.

Abbreviations: K – knit plain; p – purl; st – stitch; g st – garter stitch; st st – stocking stitch; ins – inches; cms – centimetres.

Note: The instructions are given for the 34-in (86-cm) size. Where they vary, work the figure in the first bracket for the 36-in (91-cm) size and the figure in the second bracket for the 38-in (97-cm) size.

Use No. 7 ($4\frac{1}{2}$ mm) needles.

THE BACK

Using No. 7 ($4\frac{1}{2}$ mm) needles cast on 86 (90) (94) sts and work 19 rows in g st.

Change to st st and continue until work measures 18 ins (46 cms) from beginning, ending with a p row.

For sleeves: Cast on 59 sts at the beginning of next two rows – 204 (208) (212) sts.

Continue on these sts until work measures 9 (9) ($9\frac{1}{2}$) ins 23 (23) (24) cms from commencement of sleeve, ending with a p row.

Next row: Cast off 83 (85) (87) sts, k 38, including st left after casting off. Cast off remaining 83 (85) (87) sts. Leave the centre 38 sts on a spare needle.

POCKET BACK

Using No. 7 ($4\frac{1}{2}$ mm) needles, cast on 60 sts. Work 9 rows in st st, ending with a k row. Leave these sts on a spare needle.

THE FRONT

Using No. 7 ($4\frac{1}{2}$ mm) needles, cast on 86 (90) (94) and work 19 rows in g st.

Pocket row: Work in st st. Knit 73 (75) (77) sts, slip last 60 sts just worked onto a spare needle. K to end of row.

Next row: P 13 (15) (17) sts, p across the 60 sts of pocket back, p to end of row.

Work 33 rows in st st ending with a k row.

Next row: P 13 (15) (17) sts, cast off 60 sts, p to end. Break off wool.

Return to 60 sts left on spare needle and with wrong side of work facing rejoin wool and proceed as follows using No. 7 ($4\frac{1}{2}$ mm) needles:

1st row: K 5, p 50, k 5 sts.

2nd row: K 60 sts.

Repeat these 2 rows 16 times and then the 1st row once. Break off wool.

Return to main part of front and, with right side facing, rejoin wool at side edge.

Next row: K 13 (15) (17) sts, k 60 sts from pocket front, k 13 (15) (17) sts.

Continue in st st until work measures 16 ins (46 cms) from beginning, ending with a p row.

Next row: All k.

Next row: P 27 (29) (31) sts, k 32 sts, p 27 (29) (31) sts.

Next row: K 86 (90) (94).

Repeat last 2 rows 3 times more.

Next row: P 27 (29) (31) sts, k 32 sts, p 27 (29) (31) sts.

To divide for neck opening:

Next row: K 43 (45) (47) sts, turn, slipping remaining sts onto spare needle.

Next row: K 16 sts, p to end.

FOR SLEEVES

Cast on 59 sts [102 (104) (108) sts].

Continue on these sts, working in st st but knitting the 16 sts at the edge of the neck on every row until the front measures the same as the back ending at the side edge.

Cast off 86 (88) (90) sts, k to end. Leave remaining 16 sts on spare needle.

Join in wool at neck edge of remaining sts, k to end and complete to match other side, casting on 59 sts at the beginning of the following row. Reverse shapings where necessary and maintain 16 sts in g st at neck edge.

THE HOOD

(Worked in g st)

First join shoulder seams.

With right side of work facing, rejoin wool and using No. 7 ($4\frac{1}{2}$ mm) needles k 16 sts from spare needle on right side of neck, k across 38 sts of back neck and the 16 sts of left side of neck (70 sts).

Next row: All k.

Next row: Increase 30 sts evenly across row (100 sts).

Next row: All k.

Next row: K 50 sts, turn, slipping remaining sts on spare needle.

K until work measures 13 ins (33 cms) from beginning of hood, ending at back edge.

Cast off 4 sts at beginning of next and every following alternate row until 30 sts remain.

Cast off. Rejoin wool to remaining 50 sts and work a second side to match.

The sleeve edging: Using No. 7 ($4\frac{1}{2}$ mm) needles, pick up sts evenly around edge of sleeve and work 19 rows in g st. Cast off.

TO MAKE UP

Join underarm sleeve seams. Join top and back seam of hood. Turn back the edge of the hood by 2 in (5 cm) and catch down. Stitch down bottom of pocket lining.

Make four ties from twisted strands of wool and attach to front opening.

Press lightly.

Above: Wool in cream and shades of grey can all be spun from a single piebald fleece of the Jacob

Crew Neck Sweater

Measurements
given in inches
(centimetres in brackets)

To fit chest size

38 (97)	40 (102)	42 (107

Length to shoulder

27½ (70)	27½ (70)	27½ (70)

Sleeve seam

18 (46)	18 (46)	18 (46)

Weight of wool: 12 oz (340 g)

Tension: Work at a tension of 10 stitches and 14 rows to 2 × 2 ins (5 × 5 cms) measured over stocking stitch using No. 8 (4 mm) needles.

Abbreviations: K – knit plain; p – purl; st – stitch; inc. – increase; dec. – decrease; st st – stocking stitch; ins – inches; cms – centimetres.

Note: The instructions are given for the 38-in (97-cm) size. Where they vary, work the figure in the first bracket for the 40-in (102-cm) size and the figure in the second bracket for the 42-in (107-cm) size.

Use one pair of No. 10 (3¼ mm) and one pair of No. 8 (4 mm) needles.

THE BACK

Using No. 10 (3¼ mm) needles, cast on 122 (126) (130) sts.

1st row: K 1, p 1, to end of row.

2nd row: P 1, k 1, to end of row.

Repeat last 2 rows until work measures 2¾ ins (7 cms).

Change to No. 8 (4 mm) needles and work in st st until back measures 18 ins (46 cms), ending with right side facing.

To shape armholes: Cast off 4 sts at beginning of next 2 rows, then dec. 1 st at each end of next and every alternate row until 86 (90) (94) sts remain. Work until back measures 27½ ins (70 cms) from beginning.

To shape shoulders: Cast off 6 (8) (10) sts at beginning of next 2 rows, then 8 sts at beginning of next 4 rows. Leave remaining 42 sts on a spare needle.

THE FRONT

Work as for back until the end of the armhole shaping, then work straight until front measures 25 ins (64 cms) ending with right side facing.

To divide for neck: K 31 (33) (35) sts. Turn, and leave remaining sts on spare needle.

Continue on these sts, dec. 1 st at neck edge on every row until 22 (24) (26) sts remain. Work straight until front exactly matches back at armhole edge.

To shape shoulders: Cast off 6 (8) (10) sts at beginning of next row and 8 sts at beginning of two subsequent alternate rows.

Slip centre 24 sts onto spare needle. Rejoin wool to inner end of remaining 31 (33) (35) sts. Complete to match first side.

SLEEVES

Using No. 10 (3¼ mm) needles, cast on 52 (54) (56) sts. Work 2¾ ins (7 cms) in k 1, p 1 rib.

Shape sleeves by inc. 1 st at each end of 6th and every following 7th row until there are 84 (86) (88) sts.

Work straight until sleeve seam measures 18 ins (46 cms), ending with right side facing.

To shape top: Cast off 4 sts at beginning of next 2 rows, then dec. 1 st at each end of next and every k row until 40 (42) (44) sts remain.

Dec. 1 st at each end of every row until 20 (22) (24) sts remain. Cast off.

NECKBAND

Join right shoulder seam. Using No. 10 (3¼ mm) needles pick up and k 22 sts down left side of neck, 24 sts from centre, 22 sts from right side of neck and 42 sts from back. Work 10 rows in single rib. Cast off loosely.

TO MAKE UP

Press lightly. Join left shoulder and neckband seam. Join side and sleeve seams. Set in sleeves.

Above: The sweater knitted with Suffolk wool

Opposite: A striped effect is obtained using dark Jacob wool

Ribbed Scarf

Weight of wool: 7 oz (200 g)

Tension: Work at a tension of 9 stitches and 12 rows to 2 in (5 cm) measured over stocking stitch using No. 7 ($4\frac{1}{2}$ mm) needles.

Abbreviations: K – knit plain; p – purl; st – stitch.

Use a pair of No. 7 ($4\frac{1}{2}$ mm) needles.

METHOD
Using No. 7 ($4\frac{1}{2}$ mm) needles, cast on 48 sts and work in single rib (k 1, p 1) until the scarf is the desired length.

Cast off loosely.

TO FINISH
Attach fringes evenly across the cast on and the cast off edges.

FRINGES
Make the fringes by winding wool around a suitably sized book or piece of stiff card and then cut the strands where they pass around one edge of the book. Attach these (still doubled) lengths in groups to the scarf with a crochet hook.

See page 136.

Long scarves may be knitted in single, double or treble rib.

Natural dark and light wool from Jacob sheep are plyed together and knitted up to make this scarf

GLOVES

Tension: Work at a tension of 11 stitches and 15 rows to 2 × 2 ins (5 × 5 cms) measured over stocking stitch using No. 9 ($3\frac{1}{2}$ mm) needles.

Weight of wool: $1\frac{1}{2}$ oz (40 g)

Abbreviations: K – knit plain; p – purl; st – stitch; st st – stocking stitch; m – make a stitch; tog. – together; ins – inches; cms – centimetres.

Use a pair of No. 9 ($3\frac{1}{2}$ mm) needles.

METHOD

Using No. 9 ($3\frac{1}{2}$ mm) needles cast on 38 sts, work 16 rows in k 1, p 1 rib.

Next row (wrong side): K 1, p to last st, k 1.

1st row: All k.

2nd row: K 1, p to last st, k 1.

3rd row: K 20, m 1, k 2, m 1, k to end.

Work 3 rows straight in st st.

7th row: K 20, m 1, k 4, m 1, k to end.

Work 3 rows straight in st st.

11th row: K 20, m 1, k 6, m 1, k to end.

Work 3 rows straight in st st.

15th row: K 20, m 1, k 8, m 1, k to end (46 sts).

Work 3 rows straight in st st.

THUMB

1st row: K 30, turn.

2nd row: K 1, p 9, cast on 3 sts, turn.

Work 12 rows in st st on these 13 sts.

To shape top:

Next row: (K 1, k 2 tog.) 4 times, k 1.

2nd row: K 1, p to last st, k 1.

3rd row: (K 1, k 2 tog.) 3 times.

Break yarn, thread through remaining stitches, draw up and fasten off securely. Sew up thumb seam.

With right side of work facing, rejoin yarn, k 16 and pick up and k 4 sts from cast on sts at base of thumb. K to end (40 sts). Work 11 rows.

FIRST FINGER

K 26, turn.

Next row: K 1, p 10, cast on 2 sts.

Work 14 rows in st st on these 13 sts.

To shape top:

1st row: (K 1, k 2 tog.) 4 times, k 1.

2nd row: K 1, p to last st, k 1.

3rd row: (K 1, k 2 tog.) 3 times.

Complete as for thumb.

SECOND FINGER

With right side facing, k up to 2 sts from cast on sts at base of 1 st finger, k 5, turn.

Next row: K 1, p 11, cast on 2 sts, turn.

Work 16 rows on these 14 sts.

To shape top:

1st row: (K 1, k 2 tog.) 4 times, k 2.

2nd row: K 1, p to last st, k 1.

3rd row: (K 1, k 2 tog.) 3 times, k 1.

Complete as for thumb.

THIRD FINGER

With right side facing, k up to 2 sts from cast on sts at base of 2nd finger, k 5, turn.

Next row: K 1, p 11, cast on 2 sts, turn.

Work 14 rows on these 14 sts.

Shape top as for 2nd finger.

Complete as for thumb.

FOURTH FINGER

With right side facing, k up to 2 sts from cast on sts at base of 3rd finger, k to end.

Next row: K 1, p 9, k 1.

Work 12 rows on these 11 sts.

To shape top:

1st row: (K 1, k 2 tog.) 3 times, k 2.

2nd row: K 1, purl to last st, k 1.

3rd row: (K 1, k 2 tog.) twice, k 2.

Complete as for thumb, stitching seam to end of glove.

Make left glove, reversing all shapings.

Above: Attractive designs can be created with different coloured wools using this plain glove pattern

Triangular Crocheted Shawl

Weight of wool: $4\frac{1}{2}$ oz (130 g)

Abbreviations: Ch – chain; ss – slip stitch; trbl – treble; dc – double crochet; sp – space.

Use No. 2 (7 mm) crochet hook.

METHOD

1st row: 4 ch. Into 4th ch from hook, work 2 trbls. Turn.
2nd row: 3 ch. Into the space between the first and 2nd trbl of previous row, work 2 trbls. 1 ch. Into top of 2nd trbl of previous row, work 3 trbls. 1 ch. 3 trbls. 1 ch. Into space between 2nd and 3rd trbls of previous row, work 3 trbls. Turn.
Next row: 3 ch. Into last trbl of previous row, work 2 trbls. 1 ch. Into 1st ch sp of previous row, work 3 trbls. 1 ch.

Continue working 3 trbls, 1 ch into each ch sp of previous row, keeping these sts (3 trbls 1 ch 3 trbls) constant at the point of the triangle. Continue until shawl is the desired length.

Attach fringes to the two shorter sides of the shawl. See page 136.

Above: Jacob wool shawl

Tabard

Measurements
given in inches
(centimetres in brackets)
To fit bust size
34 to 36 (86 to 91)
Length
28 (70)

Weight of wool: 9 oz (225 g)

Tension: Work at a tension of 20 stitches and 34 rows to 4 × 4 ins (10 × 10 cms) measured over single moss stitch using No. 8 (4 mm) needles.

Abbreviations: K – knit plain; p – purl; st – stitch; ins – inches; cms – centimetres.

Use No. 8 (4 mm) needles.

THE BACK

Using No. 8 (4 mm) needles cast on 110 sts and work 28 ins (70 cms) in moss stitch as follows:
1st row: K 1, p 1, repeat to end of row.
2nd row: K 1, *k 1, p 1; repeat from * to end of row.
To divide for fronts: Keeping moss stitch correct, work and place 43 sts stitches on a spare needle, cast off centre 24 sts, work to end of row.

LEFT FRONT

On last 43 sts, continue working in moss stitch until front measures 30 ins (75 cms). Cast off in pattern.

RIGHT FRONT

Rejoin wool to inner end of remaining 43 sts and work to match the left front.

TO MAKE UP

Press lightly. Make a belt and four ties for each side of the tabard, using twisted strands of wool.

Bobble Hat

Measurement: to fit an average size head.

Weight of wool: 3 oz (85 g)

Tension: Work at a tension of 11 stitches and 6 rows to 2 inches (5 cms) measured over stocking stitch using No. 5 ($5\frac{1}{2}$ mm) needles.

Abbreviations: K – knit plain; p – purl; st – stitch; tog. – together; ins – inches; cms – centimetres.

Use No. 5 ($5\frac{1}{2}$ mm) needles.

METHOD

Using No. 5 ($5\frac{1}{2}$ mm) needles cast on 60 sts and work in treble rib (k 3, p 3) for 9 ins (23 cms).

Decrease for top of hat

Next row: *K 2 tog., k 1, p 3, k 2 tog., k 1, p 3. Repeat from * to end of row (50 sts).

Next row: *K 3, p 2, k 3, p 2. Repeat from * to end of row.

Next row: *K 2, p 2 tog., p 1, k 2, p 2 tog., p 1. Repeat from * to end of row (40 sts).

Next row: K 2, p 2; repeat to end of row.

Work two more rows in k 2, p 2 rib.

TO FINISH

Break yarn 6 ins (15 cms) from needle and then thread back through sts on needle, draw work up and fasten off securely. Make a bobble and sew it to the top.

Above: Fleece from Black Suffolk sheep was spun with slub to give a texture to this bobble hat

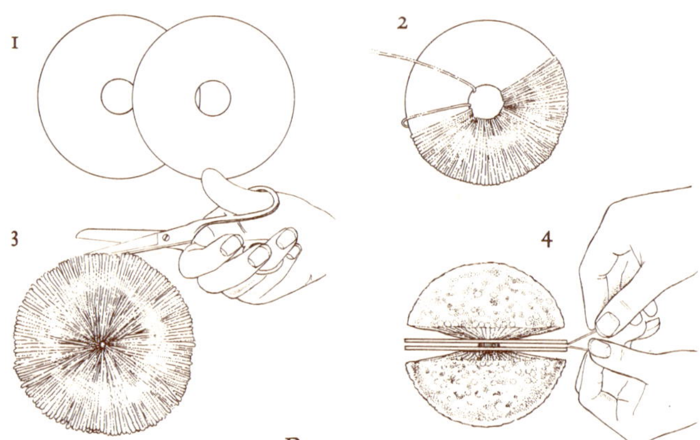

Bobble

1. Cut two circles of cardboard which dictate the size of the finished bobble, and cut a hole in the centre of each
2. Wind wool evenly round both the circles of card until the hole is filled in
3. Cut through the wool round the outer edge and between the cardboard circles
4. Tie a piece of yarn between the card circles and draw it very tight to form the bobble

Crocheted Scarf

Weight of wool: 2 oz (55 g)

Abbreviations: Ch – chain; ss – slip stitch; trbl – treble; dc – double crochet; sp – space.

Use No. 2 (7 mm) crochet hook.

METHOD

Make a ch of 36 sts.

1st row: Into 3rd ch from hook, work 3 trbls. *Miss 2 ch. 1 dc into next ch. 4 ch. Miss 2 ch. 3 trbls into next ch. Repeat from * to end, ending with 1 dc into last st. Turn.

2nd row: 3 ch. Into dc of previous row, work 2 trbls. *1 dc into 4 ch space of previous row. 4 ch. 3 trbls into next dc of previous row. Repeat from * to end, ending with 1 dc into 1st trbl of previous row. Turn.

Repeat row 2 until scarf measures required length.

Attach fringes to both ends of scarf. See page 136.

Above: Finely-spun Jacob yarn was used to make this scarf

Crocheted Stole

Weight of wool: 3 oz (85 g)

Abbreviations: Ch – chain; ss – slip stitch; trbl – treble; dc – double crochet; sp – space.

Use No. 2 (7 mm) crochet hook.

METHOD

Make a ch of 98 sts.

1st row: Into 3rd ch from hook, work 2 trbls. Miss 2 ch. 1 dc into next ch. 4 ch. Miss 2 ch. 3 trbls into next ch. Miss 2 ch. 1 dc into next ch. *4 ch. Miss 2 ch. 3 trbls into next ch. Miss 2 ch. 1 dc into next ch. Repeat from * to end.

2nd row: Turn. 3 ch. Into dc of previous row, work 2 trbls. *1 dc into next 4 ch sp., 4 ch, 3 trbls into dc. Repeat from * to end, 1 dc into 1st trbl of previous row.

Repeat 2nd row to desired length [92 to 120 cms (3 to 4 ft)].

Attach fringes to 3 sides of stole. See page 136.

Above: Natural Suffolk stole

Opposite: Natural dyed and undyed Jacob wool makes this cushion cover

Cushion

The cushion may be made using odd lengths of spun yarn and is therefore useful for using up short lengths of both dyed and undyed wool.

It is made to the basic granny square design and any of the stitch variations described below may be used to increase the size of the square.

The various colours may be joined at any point.

Weight of wool: 1 oz (25 g)

Abbreviations: Ch – chain; ss – slip stitch; trbl – treble; dc – double crochet; sp – space.

Use No. 2 (7 mm) crochet hook.

1st round: Make 4 ch. Join with ss to form a circle.
2nd round: 2 ch, 2 trbls, 1 ch, 3 trbls. *3 trbls, 1 ch, 3 trbls. Repeat from * twice. Join with a ss to top of first trbl.
3rd round: Slip st to next 1 ch sp. 2 ch. Into one ch sp just made, work 2 trbls, 1 ch, 3 trbls, 2 ch.

Into next ch sp, work 3 trbls, 1 ch, 3 trbls, 2 ch. Repeat into each corner. Join with ss to top of 1st trbl. Turn.
4th round: 2 ch. Into 2 ch sp just made, work 2 trbls, 2 ch. Into each corner ch sp, work 3 trbls, 1 ch, 3 trbls. 2 ch. Into each 2 ch sp, work 3 trbls, 2 ch.

Join with ss into top of first trbl.
5th round: Into top of each trbl, work 1 dc. Into each corner ch sp, work 3 dcs. Into each 2 ch sp, work 2 dcs. Ss to first dc. Fasten off.
6th round: Taking each side of the square separately, work 4 triangles as follows: rejoin wool at a corner, along first side, work 1 dc into each dc of the previous row. Turn.

Decrease 1 dc at each end of this row and every following row until only first remains. Fasten off.

Repeat this triangle formation three more times until all four sides are completed.
7th round: Working around the square again, as before, work 3 dcs into each corner. Into each row end (of triangle) work 1 dc. Into sp between row ends, work 1 trbl. Work along all four sides.

Join with ss to first dc.
8th round: Work a round of dc, working 3 dcs into each corner sp.
9th round: Work a triangle on each side of the square as in round 6.
10th round: As 7th.
11th round: As 8th.
12th round: Work a round of ½ trbls, working 3½ trbls into each corner sp.
13th round: Work a round of trbls, working 3 trbls into each corner sp.
14th round: As 8th round.
15th round: As 13th round.
16th round: As 8th round.

Any combination of the above sts may be used to increase the size of the cushion. Work until the desired size is reached.

To complete the cushion, a piece of hessian measuring 1 in (2.5 cms) larger than the crocheted piece all round is required.

Make a ½-in (12-mm) hem all round the hessian and attach it to the crocheted piece on three sides. Insert a covered cushion of suitable size and oversew the fourth side.

When the cushion is complete, the crocheted side will stretch to the dimensions of the hessian. Fringes can be attached to the edges of the cushion with a crochet hook.

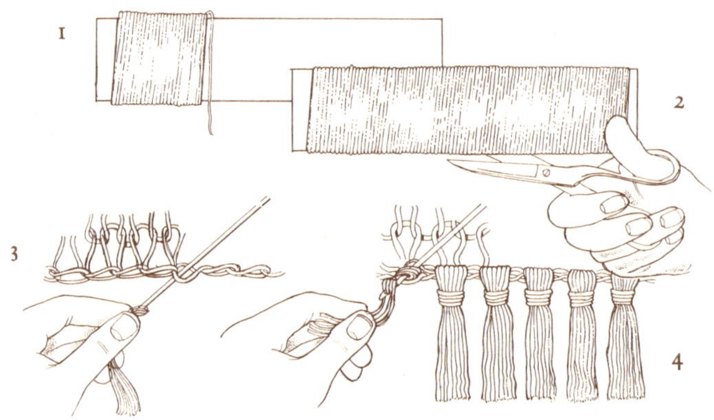

Fringe

1. Take a length of stiff card about 10 cm (4 in) wide, and wind the wool round it
2. Cut through wool along one edge
3. Take three or four strands and pull them through a hole at the edge of the cushion with a crochet hook
4. Pull the rest of the wool strands through the loop. Repeat this process along the edge of the scarf until the fringe is complete

List of Suppliers

Creative Crafts,
Winchester Road,
Chanders Ford,
Hampshire SO5 2ZG
spinning wheels, accessories, fleeces
open normal shop hours

Handweavers Studio & Gallery Ltd.,
29 Haroldstone Road,
Walthamstow,
London E17 7AN
spinning wheels, accessories, fleeces
open normal shop hours, closed Mondays

Frank Herring & Sons,
27 High West Street,
Dorchester,
Dorset DT1 1UP
spinning wheels, accessories, fleeces
open normal shop hours

Eliza Leadbeater,
Rookery Cottage,
Dalefords Lane,
Whitegate,
Northwich,
Cheshire CW8 2BN
spinning wheels, accessories, fleeces, natural dyes
by appointment

Matheson Dyes,
Marcon Place,
Hackney,
London E8 1 LP
natural dyes
by appointment

Muswell Hill Weavers,
65 Rosebury Road,
London N10 2LE
natural dyes
by appointment

Olicana Crafts,
27 The Colonnade,
Piece Hall,
Halifax,
West Yorkshire
spinning wheels, accessories
open Thursday, Friday and Saturday

The Strangers Wool Shop,
5 Timber Hill,
Norwich,
Norfolk
spinning wheels, accessories, fleeces, natural dyes
open normal shop hours

British Wool Marketing Board,
Oak Mills, Station Road,
Clayton, Bradford BD14 6JD
fleeces produced in British Isles
mail order

Bibliography

Eliza Leadbeater
Handspinning
Studio Vista, 1976

Candace Crockett
The Complete Spinning Book
Watson Guptill, New York, 1977

Allen Fannin
Handspinning
Van Nostrand Reinhold, 1970

Patricia Baines
Spinning Wheels, Spinners and Spinning
Batsford, 1977

Molly Duncan
Creative Crafts in Wool and Flax
Bell, 1971

Index

Page numbers in italics indicate illustrations